# Amazon.com Book Reviews
## *Out of the Valley*

*Out of the Valley* is the story of a remarkable person and the role of his Christian faith in positively and profoundly touching the lives of numerous others. ...From Barry Willey's expert telling, one reads how Jon, through extraordinary will and skill, rises to meet each challenge that crosses his path and each time pulls himself up and out of a valley that would swallow many others. One is struck at how Jon builds his rising life upon the timeless Christian principles of truth, freedom, justice, and love, and is prompted to follow a similar spiritual path.
~ William

In the book *Out of the Valley*, COL (R) Barry Willey does an exceptional job of depicting how the life choices and influence of one person can impact so many others. The story of Jon Shine, an American hero who was killed in action in the jungles of Vietnam, is an inspirational story of one man who was willing to take the time to extend a spiritual hand to others. It really helped me understand that I really can make a big difference in the lives of others, just as Shine did.
~ Rob

*Amazon.com Book Reviews*

The focus of one-on-one discipling from this story, coupled with the process in my life, has shown me that the discipling relationship is a vital key to maintain growth and become a victorious Christian. This book is the story of an amazing person, Jon Shine, who left an eternal legacy which Barry Willey carried out.
~ Michael

I read this during a small group study. Barry does a great job of taking Jon's life and personally applying it to the reader. I thoroughly enjoyed going through the questions at the end of each chapter with my small group. Jon Shine, even after death, is having an impact on lives for Christ, and that is truly amazing, so why not take lessons from his life and apply them to your own.
~ Yuri

This great story provides powerful insight into living for Jesus even when it is the "hard right." The inspiring example put forth by Jon Shine's life is a daily challenge for me to live up to. More than anything else, I was struck by one simple paragraph: "by focusing on the only two things that really last... that really have eternal value... the Word of God and the souls of men." This statement hit me in a way that I did not anticipate. It distills the essence of what matters.
~ Steve

# *Out of the Valley*

## An Amazing Life Story That Can Help You Make Good Choices… and Leave an Eternal Legacy

### Colonel Barry E. Willey, United States Army (Retired)

www.InvestingThatMatters.com

Creative Team Publishing
San Diego, California

© 2015 by Barry Willey.

All rights reserved. No part of this book may be reproduced, stored in a retrieval system or transmitted in any form or by any means without the prior written permission of the publisher, except by a reviewer who may quote brief passages in a review distributed through electronic media, or printed in a newspaper, magazine or journal.

Scripture quotations marked NIV are taken from the HOLY BIBLE, NEW INTERNATIONAL VERSION® Copyright© 1973, 1978, 1984 by International Bible Society. Used by permission of Zondervan. All rights reserved.

Scripture verses marked KJV are taken from the King James Version of the Holy Bible. The KJV is in the Public Domain.

Logo Design: Brian Gundell

Author's Note: In assembling information for this book, I did much research, which included interviews of many people, mentoring sessions, Bible studies and discipleship training. In this, I performed every effort to accurately portray the context and full meaning of my quotations and references resulting from such research. However, I may have unintentionally misquoted or misrepresented a source or failed to provide its complete context and, upon notification and proof of error, will place a correction on the book and company website.

Softcover Edition

ISBN 978-0-9963719-5-7

PUBLISHED BY CREATIVE TEAM PUBLISHING
www.CreativeTeamPublishing.com
San Diego
Printed in the United States of America

# *Out of the Valley*

An Amazing Life Story
That Can Help You
Make Good Choices… and
Leave an Eternal Legacy

**Colonel Barry E. Willey,
United States Army (Retired)**

*...How is the situation up there now regarding your own spiritual life and that of the other guys (Ron, Percy, John, Barry B.)? I have had some pretty big ups and downs — the lack of a definite schedule and regular study and fellowship has been pretty hard for me to adjust to — but the Lord has, as usual, been faithful and I feel like I'm starting up out of a valley again.*

*I'm continually impressed that we need Him as the center of our lives so much that not only do our consciences bother us, but we feel almost totally psychologically wrong when we're out of tune with Jesus.*

~ 2LT Jon Shine in a letter to Cadet Barry Willey, 18 Nov 1969

# Dedication

This book is dedicated to my precious Barb, Rachael, Jonathan, Jme, David, and Michael.
You are the world to me…
And to my father and mother, who loved me, invested in me, modeled Christ in all they did, and left an eternal legacy—we'll see you soon, Mom and Dad.

# Table of Contents

**Foreword**     17
    Bruce L. Fister
       Lieutenant General, U.S. Air Force (Retired)
       Executive Director, Officers'
          Christian Fellowship

**Preface**     19

*****

**Chapter One**

    It Starts with an Oath     21

    Roll-up     33

**Chapter Two**

    Firstie Year     37

    Roll-up     57

## Chapter Three

    The Real Army    63

    Roll-up    81

## Chapter Four

    The Battle    85

    Roll-up    100

## Chapter Five

    An Eternal Legacy    105

    Roll-up    132

*****

## Appendix 1

    Discipleship Plan Outline    139

*****

# Appendix 2

Testimonials 147

Acknowledgements 181

Biographical Information
    Colonel Barry Willey, U.S. Army (Retired) 185

Resources 189

Products and Services 191

*****

# Foreword

Bruce L. Fister
Lieutenant General, U.S. Air Force (Retired)
Executive Director,
Officers' Christian Fellowship

The Vietnam War was a challenging time in our nation's history and has touched millions of Americans in powerful ways. We lost some of our nation's most promising young men and women, and others had their lives disrupted and changed forever. Most of us view those losses as tragedies and move on. However, one young man's death in action in Vietnam did not just stop there. His brief twenty-three years on this earth had an eternal impact on many others and serves as an inspiration to us all.

Jonathan Shine's life was cut short as he courageously led his rifle platoon against an enemy position in October of 1970. But the spiritual impact he has had on many lives has established a legacy that continues to this day.

Barry Willey has written a powerful testimony of Jon Shine's life and legacy. Barry describes the consistent life Jon lived as a West Point cadet and as a U.S. Army officer. The choices Jon made, and the Christ-centered principles by which he lived, are a model for all of us to emulate.

*Foreword by Bruce L. Fister*

Jesus Christ told His disciples as He was preparing to leave them to go to His heavenly Father, "Go and make disciples of all nations..." (Matthew 28:19 NIV). This is also our mission. As disciples of Christ, we need to be about the business of making other disciples for the Lord and His kingdom. *Out of the Valley* provides some amazing insights into a life well-lived and into the principles of investing spiritually in others. Jon Shine's investment in Barry's life, and many others, demonstrates clearly the mission Jesus gave each one of us and provides a model to follow.

My prayer is that many will read this wonderful story of a brief but inspirational life lived to exalt Christ and build His kingdom. After you have read it, use it to build into and invest in others for the sake of the high calling of Jesus Christ.

~ Bruce L. Fister

# Preface

The story of Jonathan Cameron Shine is about the amazing power of one life lived in harmony with God's eternal purpose. How can anyone determine the influence and impact of one life on those around him? Is there hard evidence that can be scientifically measured? Probably not. Most of us just press on and never realize how we affect others. There is, however, the testimony of changed lives... lives with purpose and direction... lives motivated, inspired, and dedicated to a single cause—serving Jesus Christ. Jon Shine epitomized the servant-leader and kept Jesus at the center of his life; and the choices he made reflected that focus on Christ. How did he keep the Lord on the throne and turn to Him for all his decisions? His life has a thread we'll trace that looks a lot like the phrase from the West Point Cadet Prayer— "...choosing the harder right instead of the easier wrong."

We'll explore that amazing phenomenon and travel with him on an incredible journey of life choices...right up to the heroic final choice he made on a tragic battlefield in South Vietnam. We'll also look at the way Jon invested himself in others, following his Savior's directive to "make disciples" (Matthew 28:19, NIV) and St. Paul's admonition to his charge, Timothy, "the things you have heard me say in the presence of many witnesses entrust to reliable men who will also be qualified to teach others." (2 Timothy 2:2 NIV) Jon had a gift of communicating effectively and developing powerful relationships during his life, and he used that gift to invest in others for eternity. The testimonies and lives of the men and women who knew Jon—or those who simply knew of him and were inspired by his life—are powerful evidence that one life, even one that lasted a brief twenty-three years, can change the course of many others, give those lives purpose and direction, and mightily impact succeeding generations of believers in the Lord Jesus Christ. They are Jon Shine's spiritual legacy and will carry on his work until we all meet some day in heaven. What a glorious reunion that will be!

The reader may sense that Jon Shine appears to be too perfect...always succeeding, never frustrated or disappointed...never challenged. Absolutely not so! He faced valleys and tough challenges all the time, and those will become clear in reading his story. He was inspired and motivated by God, however, to use the gifts and talents he received and to apply them to achieve his maximum potential as a human being, West Point cadet, and U.S. Army officer. In a culture fraught with examples of lives torn by bad decisions, we desperately need an example of a real person who actively attempted—and succeeded—in following Christ's example and modeling that consistently for others. He wasn't perfect, for sure, and this narrative makes that clear. But Jon Shine was unique! That's why his story is worth telling...and his life worth emulating.

# Chapter One
# It Starts with an Oath

On the first day of July in 1965, Jon Shine took the Oath of Allegiance at West Point and became a Cadet Fourth Classman or West Point Plebe. He swore that day to "support the Constitution of the United States, and bear true allegiance to the National Government ...to maintain and defend the sovereignty of the United States." For the next four years as a cadet, and for a year as an officer, Jon Shine kept that oath. The journey to the ultimate consummation of that oath for Jon is powerful, emotional, spiritual, and poignant. Let's begin that journey.

New Cadet Shine had read about the cadet honor code in the West Point catalog sent to all applicants of the academy. He had also heard about it from his older brother, Al, who had survived the ordeal of cadet life and graduated in 1963. Their parents had raised them by a strong moral code of love and integrity, and they had had witnessed the ethical conduct of their father, a former Army officer and now business executive. Jon was now ready, even eager, to take on the challenges of cadet life.

*Jon Shine's senior high school picture*

West Point life is foremost about leadership. The first paragraph in the section of cadet regulations on leadership states a well-known maxim: "Leaders are not born; they are made." It further states that a cadet must have an active rather than a passive approach to leadership development. Jon was about to embark on the experience of a lifetime, with challenges that would stagger most people his age.

Those challenges and valleys would send him into physical, spiritual, moral, and emotional valley after valley, but he found creative ways to stay Christ-centered and climb out of those valleys, usually helping others do the same along the way.

## The Fourth Class System

*Jon, in training to be a sport parachutist in 1964, while a high school student and before heading to West Point*

What was Jon getting into when he took the oath on that first day as a cadet? It's called the Fourth Class System, an all-consuming lifestyle like few in any other institution in the world.

The Fourth Class System is designed to help cadets begin their physical, mental, and psychological conditioning to eventually graduate into the ranks of the regular United States Army as officers. It can even be viewed as a "weeding out" process for those not cut out for the challenges of cadet or Army life.

Those who have not experienced the rigors of the Fourth Class System, or cadet life in general, might mistake a West Pointer's pride for boastful arrogance. For some, unfortunately, that may be the case. For most, however, there is truly a bond that develops between graduates that lasts a lifetime. It transcends the pettiness of life's problems. It goes deep, to the heart and soul of being brothers and sisters in a profession that calls upon them to wake up each day, look in a mirror, and decide that today they are willing to give their lives for their country—and more importantly, for their buddies—should that sacrifice be required.

While the Fourth Class System has evolved over more than 200 years of the academy's existence, life as a plebe is not fun. In July of 1965, when Jon Shine arrived, hazing—the brutal, demeaning activity that surrounds being the new guy in a system built on tradition—was alive and well. The pressures Jon felt that first day included dozens, even hundreds, of Second and First Classmen (juniors and seniors) yelling in his face and shouting orders that didn't mean much to a plebe. It was just noise...loud, obnoxious noise.

Barbara Tuchman, Pulitzer Prize winner and author of the biography of West Point Class of 1904 graduate General Joseph Stillwell, very artfully describes the plebe experience by quoting Stillwell from his diary: "Brace, brace, brace," [he] wrote, "drill, drill, drill; Oh, Lord... Sink, setting up drill, drink, rest, squad drill, dinner, clean guns, squad drill, retreat, company drill around the area before supper...taps, oblivion, reveille at 4:30, brace all the time, at meals between every mouthful, had to brace on toes for an hour-and-a-half once."[1]

---

[1] *Stillwell and the American Experience in China, 1911-45*, The MacMillan Co., New York, 1970, p13.

New Cadet Jon Shine experienced the blur of "R" Day or Reception Day, survived it like thousands before him, and was assigned to a room with two roommates, for their first night of "Beast." To Dave Jamison, his new roommate from Arkansas, Jon Shine was "the first person I talked to 'as an equal' that night." Dave was overwhelmed and confused by the craziness and chaos of that day and could only think about why he had gotten himself into this mess. Ready to quit then and there, Dave wasn't sure about this new guy.

If Jon was like most every other new cadet to endure the rigors of Beast Barracks, he felt like quitting any number of times that first day and probably almost every day in that six-week valley of pain and torture. His outward demeanor didn't show it, though. He chose to keep a positive attitude and work through each challenge. His penchant for helping others soon became evident.

"Jon warmly introduced himself saying something like 'we can make it if we work together,'" Dave remembered. Here was a true friend that Dave would lean on for the next four years.

Many people who knew Jon said he had a photographic memory which helped him thrive in his new cadet experience and, combined with his strong intellect, caused him to excel at all his academic pursuits. Dave Jamison credited Jon's amazing memory—and his willingness to share it—for the success he had learning and "spouting off" the myriad of traditional "plebe poop" (traditions and customs of the West Point experience) that was thrust upon all new cadets. The intent behind the memory drills was to inculcate academy ideals and traditions, while applying pressure to the aspiring officer-candidate by all means short of placing him in actual combat. That meant correction, physical activity to the point of exhaustion, and rote memorization drills that would inevitably bring failure and then more correction, including lots of "bracing." Bracing involved the plebe compressing his chin back into his throat while keeping his head erect, creating many wrinkles between chin and Adam's apple in the process. An upperclassman would often demand from a plebe that a specific number of wrinkles be produced in the course of a bracing drill. He would then proceed to count the wrinkles and if insufficient numbers were counted, more bracing. Hazing and bracing were demeaning and hardly positive or practical leadership tools that were eventually eliminated from the repertoire of upper-class options. It is a matter of bragging rights, however, for West Point classes that endured those traditions for an entire year.

## The Harder Right

Learning to deal with the pressure was one of the goals of the Fourth Class System, and the upperclassmen were very good at dishing it out. Of course they had all lived through it and were intent on making each successive class's experience even harder than theirs.

Dealing with the daily pressures placed upon a plebe by his seniors was not fun and became a constant thorn in the plebe's side. Jon Shine was tall and lanky, and made an easy target for upper-class hazing. Dave Jamison, Jon's Beast Barracks roommate, described Jon as confident, with a positive attitude and willingness to use his talents and skills to help his buddies. He tells it like this:

> Jon's attitude was clearly one of his strongest attributes. He never faced a challenge that dampened his enthusiasm, and his outlook became infectious to all those around him. During the early days of Beast Barracks, memorization was a key to survival—a feat Jon mastered like no other. In anyone else, such a photographic memory would have instilled jealousy, but for Jon it garnered admiration. He memorized passages so easily that he always had time to help those of us who struggled to remember even a brief phrase. I also recall occasions when he would state something during required recitation to attract the squad leader and save my hide. He knew I was having trouble and it was no coincidence when on a few occasions my turn to recite didn't come up.

Flunking out of West Point isn't unheard of, considering the chaotic mess of competing demands on a cadet's life. The pressures of a rigorous study regime alone are incredible—but add to that the marching, ceremonies, sports, and harassment of a plebe's life—and academics can take a beating. Jon had few concerns about flunking but his friend and roommate, Dave, did. Thinking he was adequately prepared for the rigors of West Point with 26 hours of college credit under his belt, Dave found that prior college didn't make West Point academics any easier and that he needed all the help he could get. That need became obvious when he reached the cusp of failing first-semester French. He had flunked the final exam and was fortunately given the chance to retake it. It was now…do or die. Flunking it a second time meant certain expulsion. Contemplating the embarrassment, the disappointment of family and friends, the heartache of returning home empty-handed after having worked so hard to gain a congressional nomination, then an appointment, then surviving Beast Barracks… and failing academics… was overwhelming.

West Point French instructors will gladly tutor "deficient" students in the evenings and on weekends, if necessary. Jon Shine, whose language was Spanish, could easily have encouraged Dave to seek out that help, but chose instead, the much harder route… to help Dave himself. He chose to spend his own free time working with Dave. He couldn't teach him the finer points of the French language; that wasn't Jon's forte. But he could teach him how to study better for the final exam and keep his morale high, allowing him to gain the upper hand and eventually pass the test.

"French was incomprehensible and unlike anything I had previously tried to learn," Dave confessed. "Jon's guidance on how to study, as well as his faith that I could succeed, did more to help me on that second exam than any instructor's guidance could have. His willingness to help and coach me was immeasurably beneficial then and on many other occasions over the next three and one half years."

Jon's attitude and approach to life at West Point—using his talents to serve others—were unique. The solid spiritual dimension to his life further bolstered his self-confidence and gave him an inner peace and spiritual plumb line that kept him focused on service to others, while he unashamedly served his Lord. Jon daily prayed his personal prayers in his room, but would also learn another prayer, which all first-year cadets had to memorize—the Cadet Prayer. The Cadet Prayer is a powerful summation of a cadet's intention to live according to a "higher standard." One poignant phrase in that prayer is a petition for strength to "choose the harder right instead of the easier wrong, and never to be content with the half-truth when the whole can be won." Jon Shine tried to live that part of the Cadet Prayer to the fullest. His life as a cadet, and later as an Army officer, epitomized choosing "the harder right" over the easier wrong.

Attracting attention to himself, to take the heat off of his fellow classmate was a choice Jon made that was risky and much harder than choosing to remain silent—smug in his self-confidence and ability to memorize all required plebe knowledge. But he chose to help his classmate...a gutsy move, for sure. Choosing that "harder right" instead of the easier wrong started to become routine procedure for Jon but was never done for the praise of others. In fact, no one else really knew of Jon's propensity to make that harder but right call, save for those he was helping. Only after many years have passed is his story becoming known.

*Cadet Jon Shine as a high bar specialist on the West Point gymnastics team. Jon was the Eastern Collegiate High Bar champion his senior year*

## Faith Begins...and Grows

Jon's confidence and persistence allowed him to make the gymnastics team his first year, and he continued to apply his strong athletic abilities to that endeavor as a second-year cadet. He was now a high bar specialist and lettered his second year on the team. To this point in time, Jon was able to overcome the challenges and valleys he encountered routinely in cadet life by using his intellect and those traits of confidence and persistence. Then it happened. Jon encountered an academy field house maintenance man.

Hank Rhinefield, a middle-aged man, loved cadets and loved sharing his Christian faith with them. As cadets would come and go to the field house for various athletic events and team practice, Hank would "catch" them individually and gently but firmly inquire as to the beliefs of each cadet he would meet. Some would be annoyed and ignore Hank. Others were interested and listened to his stories. A few would even want the faith that Hank had and would commit their lives to Christ.

Al Shine, Jon's brother, was one of those who several years earlier was convicted in his heart that he needed to become a Christian after his encounter with Hank, but waited until he got back to his room and could study his Bible and think over the things Hank had told him. He became a Christian in the quiet of his room with a simple prayer whispered to the Lord, confessing his sin, asking for forgiveness, and inviting Jesus Christ, Savior of the world, into his life and heart. Al shared what he had experienced with Jon and encouraged him to seek Hank out when he got to the academy. Jon did just that.

Hank's approach was simple. Over the exit to the field house he placed a large sign with John 3:16 inscribed upon it. Hank would then use that verse and personalize it for each cadet he would engage. *"For God so loved Jon Shine that He gave His only begotten Son, that if Jon Shine would believe on Him, Jon Shine would not perish but have everlasting life."* (KJV, adapted)

*Jon, on the right, with older brother Al, a cadet at West Point, and his older sister Sallie*

Jon, like his older brother Al, confessed his sin to the Lord, turned from it, and accepted the free gift of eternal life by believing in Jesus Christ. Hank's enthusiasm for living as a Christian got Jon fired up; and Jon, after turning his life over to His heavenly Father, never lost that fire or desire to serve. Jon's heart was touched and his life was transformed by a humble janitor who lived out his faith on a moment-by-moment basis.

## Chapter One Roll-up

<u>Principles from Jon's Life</u>

- Don't be afraid to live your faith before others... it will make a difference.

- Putting others before yourself is the first step to a servant-leader's heart.

- If you have never made that decision to become a believer in Jesus Christ, try personalizing John 3:16, as Jon did when he encountered Hank Rhinefield, a janitor at West Point.

<u>"Harder Right" Choices Jon Made</u>

- Spent his own time to help his struggling classmates academically

- Captured upper-class cadets' attention when his classmates were being hazed

- Listened to a janitor and made a life-changing commitment to Christ

## Putting and Keeping Christ at the Center

- Jon put Christ at the center of his life when he fully realized his sin and his need there in that field house at West Point. He was determined to keep Him at the center of his life...for the rest of his life.

## Discussion Questions

- In what valley or valleys did Jon find himself, and how was he enabled to climb out?

- In what valleys do you find yourself at this time?

- What will you do to get out of those valleys?

- What kind of attitude did Jon display his first day of Beast Barracks, and how did it affect his roommate?

- What is selfless service or servant leadership, and how did Jon display that throughout his plebe year?

- How can you develop selflessness or a servant's heart?

## Jon's Spiritual Legacy*

- Roommate Dave Jamison

*This is a list of those mentioned in this chapter with whom Jon Shine had a relationship and whose lives were — either directly through Jon or indirectly through one of Jon's mentees — positively impacted by Jon's life and Christian faith.

# Chapter Two
# Firstie Year

From the challenging and unique experience of a West Point plebe year, we will skip to his equally tough and formidable First Class year. For Jon, Third Class (sophomore) and Second Class (junior) were years of professional and spiritual growth and maturity. He, and others, were mentored by faculty and staff officers who took the time and interest to invest their lives in these young believers. West Point is all about leadership development, and Jon's positive attitude and ability to handle pressure helped his tactical officers see his potential.

As First Class year arrived, Jon was chosen to be the company commander of Company G-1. In charge of about 120 cadets from all four classes, Jon was well-prepared for such a key leadership position. Starting a new academic year as the cadet-in-charge means setting disciplinary standards quickly and providing a positive role model for the younger classes, as well as for his own classmates. Leading one's peers can pose challenges that will try one's patience and sense of justice. Establishing a clear direction for the company to follow in all of its endeavors is the company commander's responsibility and Jon understood that unequivocally.

*A U.S. Army photo of Cadet First Classman Jon Shine escorting a visiting dignitary from New York City, Dr. Sylvester Carter, on the West Point grounds. The star on Jon's uniform collar indicates he was a "starman," attaining the top five percent in his class academically. The stripes indicate he was a cadet captain.*

Though First Class studies are demanding and would challenge the brightest cadets, when added to the military, athletic, and extracurricular activities that consume all cadets' lives—getting ready for graduation, thinking about choice of branch (infantry, armor, artillery, engineers, etc.), thinking about a life mate, etc.—they can become overwhelming. Jon, however, had achieved exceedingly high academic standards thus far, and would be able to adequately juggle all those duties and responsibilities and maintain his academic record.

One of Jon's extracurricular activities would be taking a lead role in the spiritual development of several plebe cadets within his company, while also providing spiritual leadership and encouragement to his classmates and fellow Christian believers throughout the Corps of Cadets. Jon met an officer during his First Class summer trip—Captain Paul Stanley—at Fort Benning, Georgia, who encouraged him to take such a key leadership role within the Corps of Cadets. Paul Stanley (a classmate of Jon's brother Al) would soon be stationed by the Army at West Point as an Admissions Officer and would become a spiritual mentor to Jon for his final year. Despite his many other activities and duties, Jon was very desirous of leading in this meaningful way—personal and corporate Christian maturity—a path he had followed faithfully since becoming a cadet and was not about to abandon now.

In a very telling letter dated April 1969 to his older brother Al, then serving in Vietnam, Jon weaves a tapestry of candor, brotherly admiration, professionalism, subtle humor, self-effacing modesty, and spiritual insights.

> *Spirit still working overtime here. Last weekend Don Moomau preached here and then spoke informally Sunday afternoon. He was All-American linebacker at UCLA in '53 and now is a minister in the LA area. His testimony and real, sincere, and honest talk was, I think, one of the best we've had this year. He was competing with Gary Puckett and the Union Gap, a singing group, for an audience and didn't fare too well, but I have found real peace in this matter. I figure that with the speakers we bring up, if we do our work well, we can just leave the rest up to the Lord.*

> *This week we have started a new Bible study over here in Companies F-1 and H-1 where I now live. We have 2 firsties, a yearling and 3 plebes...*
>
> *This weekend I am CIC [cadet-in-charge] of a Protestant retreat...up to Deer Hill in Wappinger's Falls. We have about 60 guys coming, including several on the football team. I think athletes can often have a good ministry here at Woops [slang for West Point] because most guys are so sweaty. However, I am somewhat against the emphasis I see sometimes on guys being effective because of all the neat things they do — we non-champions can be used, too.*
>
> *Psalm 27:1 (NIV): "The LORD is my light and my salvation — whom shall I fear? The LORD is the stronghold of my life — of whom shall I be afraid?"*

Jon's enthusiasm for Christ and serving Him wholeheartedly as a cadet was beginning to have a marked impact on many around him. His professionalism was evident to all, but even more evident was the joy in Christ he demonstrated. It was infectious.

## Committed to Making Disciples

Jon's interests were very broad and ranged from helping younger cadets and his own classmates become stronger in their faith to making the entire cadet system function more effectively. He was given responsibilities by senior officers, both formally and informally, that would have overwhelmed most 21-year-olds in college—valley experiences for sure—not to mention many cadets at the academy. Instead of becoming flustered and backing away from such duties, Jon worked hard to overcome the obstacles and challenges of such new assignments and get on top of them, seeing most as opportunities to learn...and impart what he had learned to others.

For a confused, lonely, and scared plebe named Barry Willey, from Indianapolis, Indiana, Cadet First Classman Jonathan C. Shine was an unlikely hero. On the last day of the transition period between Beast Barracks and the academic year, when all the upperclassmen return from their summer duties, trips, and vacations, Jon confronted me while we stood in formation ready to march to the dining hall for dinner. I stood in the rear of the company formation with several of my classmates, homesick, frightened, and not sure I was where God wanted me right now. Having survived Beast Barracks but uncertain I had what it took to endure plebe year, I watched as my company commander walked deliberately to the last rank of his company, where I stood. As he moved to a position directly in front of me, I trembled with fear. Then he spoke. His simple question to me as I stood at a stiff position of attention, chin well to the rear, was, "Cadet Willey, would you like to join me in a Bible study tonight in the company basement after duty hours?" Somewhat taken aback, but pleasantly relieved that there were other Christians within the Corps of Cadets, I responded with a quick, "Yes sir!"

That brief encounter changed my life. A relationship had begun that would last a lifetime and have a profound impact on the way I lived — as a cadet, as an officer, as a husband, and as a father. Jon approached several other plebes in our company and assembled a small group of eager men who desired to grow in their Christian faith, while they progressed in their cadet experiences.

As these men's Cadet Company Commander and knowingly taking an interest in a few lowly plebes, Jon risked the ire of his classmates and other upperclassmen, setting himself up for charges of fraternization or familiarization. Another "harder right" choice. No doubt there were those who whispered behind Jon's back and talked amongst themselves about this action. Another valley that Jon would walk through confidently. The discreet manner of his involvement in our lives, however, and Jon's own stellar reputation, created a situation that was tolerated by the members of Company G-1. There was never any pressure to participate, and Jon's leading of these Bible study and spiritual mentoring groups was personable, yet scholarly and professional. And we met after duty hours. It would have been hard for anyone to find anything worth criticizing in the arrangement, if they honestly evaluated it.

### Jon Shine's Bible Study Techniques

> Author's Note: Here is an amazing statement from Jon, written to his wife while he was in Vietnam, which shows the depth of his approach to studying the Bible and teaching its principles. It represents the kind of

teaching he did in our small group sessions. He was giving her ideas about how to teach a youth Bible study she had volunteered to lead while he was deployed.

*The progression... What does it say? What does it mean? What does it mean to me?... is, in my opinion, the best one and should be adhered to. A further question, which might prove valuable to your class later on when you are more established and familiar with each other is... What am I going to do about it? In this stage I mean that each student might choose one verse or phrase from the week's lesson which pertains particularly to him or her. Then he decides what, specifically, he intends to do about it. The next week he can report whether he was faithful to his plan, and if so, what the results were.*

*We used this format in one study at West Point and it really proved valuable to me. It proved to me the value of prayer and positive effort at living a godly life.*

*Examples of what a person might decide to do would include: 1) daily prayer about a particular problem, 2) special effort in a personal relationship, or 3) any similar type action – Honey, if you pray and plan to stick to your ideas, the Lord will really give you a fruitful class. By studying the Scripture itself, you can expect that the power of the Word will work in your group.*

Our group usually met once during the week, in the evening, down in the basement of the cadet barracks where the quiet atmosphere supported a discreet study of the Bible. Participants, including members of the company other than plebes, had to be willing to sacrifice a portion of their evening that would have otherwise been devoted to studying for the next day's academics. As it turned out that year, not one of our group suffered adversely in academics.

Our meetings were rich with teaching from God's Word, and it was very evident that Jon had a broad understanding of Scripture. But it was not just a head-knowledge. His heart was completely sold out to Christ, and the applications he made of God's Word to the real world there at West Point stuck with me and helped me get through that rough plebe year.

> *Dave Jamison, Jon's roommate, reported that he was impressed to observe Jon faithfully and regularly reading his Bible...throughout his entire cadet experience. Getting into God's Word in order to grow in one's faith is certainly a tried and true way to overcome or deal with life's toughest obstacles. It's an exciting book about real people, real problems... and real answers to our toughest challenges. It will be our anchor in the storm... and our ladder to climb when we find ourselves trudging through the dark valleys of our life.*

Jon lived his faith, and all around him saw it. I admired his faithfulness and commitment and wanted the same thing for myself. His prayer life also reflected a genuine passion to effectively communicate with our heavenly Father. He modeled the prayer life that I wanted. Jon was about the Lord's work by doing what the Lord had commanded us to do… to make disciples.

This passion for investing in others was not by happenstance. Jon willingly submitted to an intensive discipleship program as a Second Classman (junior) under the tutelage of then-Captain Paul Stanley, a staff officer in the Admission Office. Paul shares some details.

> *We began meeting cadets through my job (part of it was to stay close to cadet life… in Admissions). Al Shine (Jon's older brother) told me about Jon… and we connected. Jon knew a couple of others and within two months, I was meeting with five eager juniors and one senior. All responded to a "basic discipleship" journey. We started them in consistent quiet time and with daily journaling. We studied for four months all of the Gospel descriptions of what Jesus said it meant to be a disciple. With each part, the commitments rose.*
>
> *Soon following Christ took on a whole different meaning and further implications for each of our lives and time. They learned the Bridge (a gospel illustration) and began prayerfully sharing it. We intensely prayed for all of our mates around us and did not seek to "evangelize" them, but to love them,*

*pray for them... and watch for opportunities to serve...and be ready always to give an answer to those that asked. We also invited some to join us in an investigative Bible study through Mark. We came up with key discussion questions for each chapter. It was fun and exciting what God began to do in and through the lives of these few.*

Jon was ready to begin his own discipleship program with us eager few plebes whom he had asked to join him. We met consistently, despite a very demanding schedule of academics, athletics, and cadet duties. On one occasion, we were unable to meet due to scheduling conflicts. Jon wrote a personal note to the members apologizing. His note to me was brief and to the point, yet spoke volumes. It, in fact, is a microcosm of his life at West Point—developing relationships, meeting other's needs, displaying excellence, and unashamedly exercising faith:

> Willey, 4th Cl
> Sorry about Thurs nite—I guess we all got a little busy—let's shoot for another meeting Sun. nite—Daniel 3:17,18
>
> BEAT SPRINGFIELD Mr. Shine

The verses from Daniel 3:17, 18 (NIV): "If we are thrown into the blazing furnace, the God we serve is able to deliver us from it, and he will deliver us[a] from Your Majesty's hand. But even if he does not, we want you to know, Your Majesty, that we will not serve your gods or worship the image of gold you have set up," along with the context, describe the three Jewish lads, taken into captivity by an invading king, who refused to bow to that king's idols and defiantly worshipped their God. They were thrown into the fiery furnace but were kept safe by Jesus Christ, Himself (Daniel 3:25 NIV).

Though always a popular Sunday school story with younger ones, the trials, faith, and courage of Shadrach, Meshach, and Abednego inspired Jon in his own daily faith. He wasn't ashamed or hesitant to share that inspiration with his fellow cadets. Again, his firm stance on living unashamedly as a Christian, while teaching and training others in the faith, is another example of choosing the harder right over the easier wrong of just going with the flow and not making one's faith a lightning rod for others to criticize. By living this way, Jon was deliberately choosing to walk through the valley of real and potential criticism and mockery—not a fun trial to have to endure. Jon's commitment to this aspect of his life, however, would not allow him to equivocate on the matter. He ignored the criticism. Or he dealt with it by being the best cadet he could be. He didn't talk to us about those who were critical of his Christian stance and actions but we knew that backtalk was going on. The rumor mill was alive and active throughout the corps in these kinds of matters. Nonetheless, he always overcame it.

"Beat Springfield" was typical of Jon's spirited attitude about Army athletics, especially his beloved gymnastics team. He wasn't afraid to rally the troops.

The University of Springfield (Massachusetts) had a great gymnastics team and was one of Army's biggest rivals. Jon was extremely competitive in the sport and grew anxious each week before a meet to help his team on the high bar event, which by this time he had mastered. I was blessed to also make the Army gymnastics team for the first two years of my cadet life and, during that first year, was able to observe Jon's leadership and athletic abilities up close and personal. He was by no means perfect but spent an amazing amount of energy trying to perfect his high bar routines. It paid off. His crowning achievement in gymnastics at West Point was winning the Eastern Intercollegiate High Bar Championship his senior year.

## A Tough Predicament... and a Tougher Choice

One of the good deals for cadets, particularly Firsties, was an "away" football game. This meant that part of the Cadet Corps traveled to the game site (usually within a few hours' driving distance from West Point) on a chartered bus, attended a unit meeting, and then cheered on the Army team during the game. Firsties were privileged to travel by their own means but had to show up on time for the pre-game meeting and the game. During one particular game with Rutgers University (in New Jersey), two of Jon's company classmates chose to exercise their "First Class privileges" and get to the game on their own. Unfortunately, they were involved in an auto accident and didn't make the pre-game meeting or the game. What Jon—the company commander during this time frame—discovered in his investigation was that the accident didn't happen before the game but rather near the end. There was, therefore, no excuse for the First Classmen's absence.

At West Point, not "messing over or ratting on your buddy," especially your own classmate, was an unwritten but closely monitored rule. Expectations among peers were high. Cadet leaders were always in a potential predicament when their peers got into trouble. Peer pressure can be a cruel thing and make for rough going when the "rule" is violated. Jon knew all about the rule and had seen it in operation throughout his cadet experience. He now faced a dilemma that rivaled the toughest decisions he ever had to make in his time at the academy. Should he look away and ignore the infractions, keeping peace within the Class of '69 in Company G? Should he exercise his duty as a leader and recommend the maximum punishment, get his friends' attention and hope it set an example for others who might try to take advantage of their privileges? Was there any other option available? Four classmates were involved, and all four were injured to some extent. The violation was serious and could result in serious punishment, in the form of a ration of demerits, a load of hours to be "walked" during free time on the "cadet area"—the large open cement pad in the middle of the barracks area—and a number of months confinement to their rooms.

This kind of punishment package is called a "slug" and, in fact, hits one in the gut just like a boxer's punch, especially as a First Classman. The hope—yes, the expectation—of each participant in this incident was that Jon would overlook the violation and, in light of the circumstances, leave the cadets to their own devices this close to graduation. The Company G-1 Tactical Officer, a cigar-chomping Texas A & M graduate and Vietnam vet, insisted that Jon issue the maximum punishment for the offense—about six-months' confinement in their barracks room during free time. Jon thought long and hard and decided in favor of a compromise slug, about two months confinement and a corresponding number of demerits and hours on the area. Jon was severely criticized by those involved and by many others who felt Jon really "messed over" his classmates. The heat was on, and the ire of his classmates in G-1 was rising against him. It is not fun when you've made friendships over a three-year period that you believed to be truly lasting, and those friends turn on you when you most need their support. Into the valley...

Bill Smith, one of Jon's biggest critics while they were cadets and a participant in this incident, commented on his memories of it and of Jon:

> *Jon and I did not get along too well the last couple of years at the academy. This was a direct and proximate result of my lousy attitude and Jon's consistently high standards. Jon expected firsties to make beds and such, even when he wasn't the*

company commander. But Jon impacted everyone's life in a positive manner. At the time of the incident at the Rutgers game he was the officer on duty. I remember how really perturbed I was after Jon had written me up on an offense that resulted in a 2-month "slug" and 44 winter hours walking on the area.

The day after my punishment board Jon and I met unexpectedly on the barracks steps and he spoke to me in a friendly fashion, though not apologetically. I responded briefly and we went about our business. I realized from that encounter that Jon held no personal animus towards me. He had done what he believed his job called for. Actually the two of us argued about it in a most civil fashion prior to his actually writing up the offense, and then he took some time to think it over. He then came back to me prior to submitting it to inform me of his decision. Jon Shine was a class act.

Choices. Always hard choices. The harder right. Does that phrase from the West Point Cadet Prayer ("Make us to choose the harder right instead of the easier wrong") sound like so much pious rhetoric... or is it actually a concept that, unbeknownst to some cadets, becomes other cadets' subconscious mantra?

Jon Shine wasn't the only cadet in the Class of 1969 to try to do right and honorable things throughout life at West Point and in the Army. He was, however, when viewed in comparative terms with his classmates, by their own admission, one of the few who appeared to combine compassion for people, pure intellect, common sense, fairness, honor, humor, faith, and an extremely high sense of duty consistently throughout his four-year West Point experience.

For the third and final portion, or detail, of his First Class year, Jon was chosen to be a Battalion Commander with the cadet rank of captain. A battalion is composed of four cadet companies, like the company that Jon commanded during first detail. This selection recognized his past performance and his potential, and it gave him the opportunity to exercise all he had learned to this point in one final and exciting period of cadet life. Jon maintained his status as the preeminent high bar performer on the gymnastics team, while teaching Sunday school and chairing the Cadet Chapel Forum. All this time, Jon was also spiritually mentoring several of us plebes, while spending time with some mature Christian staff and faculty officers who helped mentor him in his faith in Christ. Teaching multitasking and effective time management is a primary thrust of West Point's program of leadership development, and most cadets' lives looked as busy and multi-faceted as Jon's. His priorities were straight, however, and his walk with Christ was evident to all.

Graduation was drawing near, and preparing for life as an Army lieutenant became the one thing on most Firsties' minds. While many consider their professional legacies — how people will remember them — when completing a watershed event in their lives, the main thing on Jon's mind at the end of his cadet experience was ensuring that the spiritual legacy of those who had gone before him was carried on at West Point by those who would come after him.

*Cadet Captain and Battalion Commander Jon Shine, third from left, with his staff during his senior year at West Point*

Jon wrote the words below on a green Department of Defense routing slip addressed to me, twelve days before graduation day for the Class of 1969.

> WILLEY, 4ᵀᴴ CLASS, CO. G-1
> *Thru M/C (message center)*
> *Let it be now and henceforth known that you will report to room 3921 at 031525June for Special Inspection. This order to be superseded only by someone with six stripes.*
> *Jonathan C. Shine*
> *Cdt CPT, 2d Bn, 1ˢᵗ Regt.*
> *Commanding* [Author's note: 3:25 p.m. on June 3]

An order to report for Special Inspection strikes fear and anxiety into any plebe, especially when it comes from the Battalion Commander. I had gotten to know Jon in a more personal way throughout the year in Jon's company and studied the Bible and prayed with him during our free time. It still was not clear, though, what Cadet Captain Shine wanted with me. The note was a mystery that had me somewhat worried.

When the day arrived for graduation, I pulled out the green routing slip and double-checked the time to report and the room number. It was time. With shoes highly spit-shined, a starch-stiff pair of white cadet trousers under a full dress gray coat—brass buttons shined to perfection—and crossed white parade belt with brass plate in the center of my chest, I was ready for the worst, most scrutinizing inspection imaginable... a Special Inspection by a Cadet Battalion Commander.

Knocking on the door, I could almost hear my knees knocking at the same time. I was nervous and sweating profusely against the high, stiff collar of my parade jacket. What was about to happen? I guessed it couldn't be any worse than what I had just finished going through the past eleven months.

The door opened and there stood Cadet First Classman Jonathan Cameron Shine, five gold stripes on his full dress coat, his red officer's sash neatly tied around his waist and his gleaming saber ready at his left side. "Come in, Cadet Willey. I'm sure you are wondering why I called you here. Well, you are surely aware of the tradition of upperclassmen recognizing plebes on graduation day. It symbolizes the break from the tough plebe year to the ranks of the upper class. It involves shaking the plebe's hand and calling him by his first name." Jon then thrust his calloused gymnast's hand toward mine and said, "Hi, Barry, I'm Jon." Hesitating, but happy and relieved, I raised my right hand and firmly grasped Jon's and our eyes met and a bond was formed that day that only the few who have experienced it can comprehend. But that wasn't the end of the Special Inspection. Jon then opened and held out his left hand, which had been grasping a dulled silver dollar, a rather old vintage coin. He explained its significance to me.

Jon had received the silver dollar from a graduating cadet when he was a cadet, at a similar recognition. The tradition behind the coin transfer was this. The coin was given to a cadet who exemplified Christian character and leadership at West Point. It symbolized the faithfulness of a generation of men who were willing to risk ridicule and perhaps spiritual persecution while living a godly life as a cadet. Being recognized not only as an upperclassman but also as a spiritual leader with responsibilities to the Lord and to his fellow cadets was a distinct honor... and an awesome charge. I felt a deep awe at this nod to my potential as a spiritual leader and a little bit of trepidation, hoping and praying that I could live up to the expectations inherent in this tradition. One more handshake and a heartfelt, manly hug sealed our friendship and bond as brothers in arms and brothers in faith.

Jon left that evening for his home in Pleasantville, New York, and a well-earned respite before the requisite military schools and training that would prepare him for a combat tour in Vietnam. He had volunteered for such duty *before* he graduated. Second Lieutenant Jon Shine, who had taken the oath of allegiance to support and defend the Constitution as a commissioned officer early that day, was now ready for the toughest valleys of his life. They would soon be upon him.

# Chapter Two Roll-up

<u>Principles from Jon's Life</u>

- Right choices are usually the toughest in life. Never back down. Stand your ground. Know what you believe. Act upon it every day.

- Always weigh the circumstances, talk to others about the situation, and ask for and trust God for wisdom.

- Invest your whole self in someone else the way someone has invested in you.

<u>"Harder Right" Choices Jon Made</u>

- To approach several younger cadets in his own company and devote himself to their spiritual development, when that is clearly not the popular thing to do

- To punish his own classmates for regulation infractions only months before their graduation, when that also was not popular with his peers

- To volunteer for combat duty in South Vietnam, before he even graduated

## Putting and Keeping Christ at the Center

- The Rutgers football game incident put much pressure on Jon to overlook his classmates' infractions. Jon was not about to do that, and his faith in Christ gave him a firm foundation upon which to make the hard choice.

- Spiritually mentoring and teaching several younger men, while being mentored by a more mature believer in the Christian faith, enabled Jon to stay sharp in his own faith and get out of the inevitable valleys that cadets find themselves in, as he helped others grow and climb out of their valleys.

- Staying committed to the Lord's work (teaching Sunday school, mentoring younger Christian cadets, and being mentored by older staff and faculty officers), while doing his cadet duties and academic responsibilities, kept Jon focused on Christ and helped keep the Lord at the center of all he did.

## Discussion Questions

- What "valleys" did Jon Shine find himself in, and how did he climb out?

- What kind of peer pressure did Jon experience in this chapter of his life?

- Has peer pressure ever challenged you? How have you dealt with it?

- What would have prompted Jon to risk the ire of his classmates and company mates to spiritually mentor several younger men in the company, while modeling Christ?

- What was Dave Smith's ultimate view of his classmate, Jon? Why did he feel that way?

- How does committing to be mentored or to mentor someone in the Christian faith change one's life from the "normal"?

<u>Jon's Spiritual Legacy</u>

- Cadet Barry Willey (author of this book)
- Cadet Ron Hawthorne (testimony follows)
- Cadet Jim Hougnon (testimony follows)

**Colonel (Retired) Ron Hawthorne, Barry Willey's roommate, a Class of 1972 academy graduate and one whom Jon Shine spiritually mentored:**

My company commander (as a plebe) was Mr. Shine... One evening, as I was one of the first cadets in dinner formation, he walked over to me... "Mr. Hawthorne, are you interested in studying the Bible?" "Yes, sir!" I don't know how he knew my name or that I might be interested in Bible study. That week we met for Bible study in the basement of Company G-1, in the locker room of the old Central Area divisions.

[Authors' note: These barracks were some of the oldest barracks at West Point and date to the days when Eisenhower, MacArthur, and Patton roamed the barracks.]

I don't remember what we studied, but I do remember being thankful to God for assigning me to G-1. My time at West Point was a time of spiritual growth and maturing. I made friends for life with whom, through the years, I have studied the Bible. The bond goes deeper than being classmates, to brothers in Christ. This is only a bit of the legacy Jon Shine left. Only God knows the full extent of it.

**Colonel (Retired) James Hougnon, a Class of 1972 academy graduate and another whom Jon Shine spiritually mentored:**

Our first meeting was brief but memorable. Jon's smile was engaging and his love for Jesus Christ evident. Jon impressed me because he excelled at everything. He was a star man (top 5 percent academically at the Academy), a cadet battalion commander, and a captain of the gymnastics team. He impressed me because of his unswerving commitment to modeling Jesus Christ in his life. He impressed me because he cared about me, a mere plebe from a small town in Colorado.

During the last third of the academic year he became battalion commander and moved into the barracks near where I lived. He immediately began an evening Bible study and I eagerly joined. This was my first experience with regular Bible study and started a weekly practice that I continue today.

[Authors' note: Jim named his son Jonathan after Jonathan Shine, as did Al Shine and Barry Willey. Those young men are all walking with Christ today.]

# Chapter Three
# The Real Army

The first stop for any infantryman on active duty is Fort Benning, Georgia. Southern Georgia was extremely hot and humid in August when Jon reported for his Infantry Officer Basic Course, Airborne School, and Ranger School. A bronze soldier, much larger than life, stands in front of Building 4, the "schoolhouse" as it is called, where all infantry officers spend about twelve weeks in rather dull classroom settings studying one of the most important curricula anywhere — the art and science of winning on the modern battlefield. Others view it as the study of managed chaos. In the end, it's war and it's hell.

*"Follow Me" statue at Fort Benning, Georgia.*
Photo Credit: Captain Vernon Robinson Jr., USA

The statue is a World War II era combat infantry soldier leaning forward toward the enemy, M-1 Garand rifle in his left hand, and his right arm in a palm-forward gesture, as if he were crying out to his men, "Follow Me" — the motto of the infantry. That figure strikes awe in most spectators; but a special, intimate bond and pride exists between a soldier and that bronze man. He represents what every infantryman wants to be able to exhibit, if he were to find himself on a foreign battleground someday — valor in the face of the enemy and competent leadership skills. They learn the latter at West Point, Reserve Officer Training Course (ROTC), Officer Candidate School (OCS), and at Fort Benning. The former is something about which one can read but it is not something that can be drilled or practiced. In combat, it is either there or it is not. Most infantrymen hope it will be there but don't dwell on it or worry about it.

Combat veteran role models, from Vietnam, were in abundance in the late '60s, and virtually all of Jon's Tactical Officers were veterans. From them, cadets heard the war stories about leading men in a desperate, courageous battle, getting wounded and seeing men die. Jon's senior Tactical Officer during his First Class year was Colonel Ralph Puckett, several times wounded in both Korea and Vietnam and twice recipient of the nation's second highest award for valor — the Distinguished Service Cross. Colonel Puckett didn't boast of his courage in battle, but all who knew him saw him as the quiet, confident, heroic model of a leader that they wanted to emulate.

*LT Jonathan C. Shine, taken in August, 1970, just before deploying to Vietnam*

Before Jon did his time as a student in the Infantry Officer Basic Course, he would complete two of the most rigorous and exciting training regimens in any military in the world, Airborne and Ranger Schools. Qualifying as a paratrooper and getting to proudly wear the silver jump wings, and then successfully enduring the hardship of the nine-week Ranger instruction, qualifying the soldier to wear the black and gold Ranger "tab," are rights of passage that forever link new airborne rangers with those from the past.

Jon was no different from most of his classmates who chose infantry as their branch. He wanted to succeed in those courses and get a good start on a career as an officer, while also getting grounded in the officer basics. He knew that the airborne and ranger courses and combat in Vietnam loomed large as valleys he had to negotiate. Learning the ropes as best he could now—at Fort Benning—might give him an edge in combat.

"He would always help others," remembers Tim Rucker, one of Jon's best friends, a roommate from the academy and Jon's "Ranger Buddy" during the Ranger Course. "He would help his buddies with their patrols and they all respected him for it."

The U.S. Army Ranger Course is the toughest training school, bar none, that the Army offers. It's not for the faint of heart. Even the most physically able candidates, like Jon and his classmates surely were, have a hard time dealing with the lack of sleep and food—the peacetime equivalent to shooting bullets at you in a combat scenario. It is a valley experience second only to actual combat.

One evening during the Fort Benning Phase of Ranger, before the dinner meal, Jon's class was lined up in front of one of the dinner lanes, a dry one, while the other lane had collected water after a recent rain and remained wet with ample mud to fulfill a pig's wildest delights. When the cadre—trainers of the Ranger course—ordered half the class to enter the muddy lane, the classmates all agreed they would all enter the muddy lane. Jon's class, as others are wont to do, used humor to get through the tough parts of this course. They moved in front of the wet lane and infuriated the cadre. When the cadre again ordered half the students to the dry lane, the class insisted as a whole that if they couldn't all go through the wet lane, then they would refuse to eat… a hunger strike if you will. The cadre quickly realized they were up against a hard-nosed class with a lot of esprit, the quality sought in Army Rangers. The cadre decided to stop fighting the class and joined them, leading them through the muddy low crawl lane. Jon's class established itself rapidly as a tough, motivated group of men who desperately wanted the coveted black and gold Ranger Tab.

The tough, first phase of the course is concluded by each student negotiating an obstacle course, the "Darby Queen" (after William O. Darby, the father of modern Rangers in World War II). It requires the student to muster the strength and gumption to gut it out — despite surviving two of the toughest weeks of their lives — in order to climb the log ladders, swing on the rope across a muddy, water-filled pit, and dangle on the cargo net. The final fling on the Darby Queen was the ultimate nightmare — the log walk and slide-for-life combination. The Ranger candidate had to climb about 20 meters straight up a pole with ladder rungs, gingerly mount another horizontal telephone pole and walk to the middle, stepping carefully over a seat, and walk to the end, all the while balancing to avoid a fall into the lake. Once at the end of the pole, the student then grasps, with what strength he has left, a metal handle and begins a slide down a rope-pulley contraption that picks up speed as it closes in on the rock on the far shore of the lake. If the student doesn't drop when told by a cadre sergeant, he will smash head-on into the boulder. Once in the water, the Darby Phase of the Ranger Course is completed.

In typical Shine style, Jon deftly climbed the pole, and using his gymnast training and form, he moved effortlessly to the step in the middle, hopped over and moved on to the slide, gracefully taking it to the drop point and executing it perfectly. Tim Rucker recalled a Ranger sergeant looking around at Jon's classmates and commenting, "That's the way you all should do it." Hard work, training, and confidence enabled Jon to climb out of this "valley" phase of Ranger training and move on to the next.

The remaining two phases of the course are tougher versions of the first, only in different locales — the mountains of north Georgia for the Mountain Phase and the swamps of northern Florida for the Jungle Phase. The patrols are tougher and longer, the food and sleep are scarcer, and the emaciated, skeletal-like forms that emerge from the swamps at the end of the training are hardly recognizable by family and friends. Guts, motivation from peer and cadre leadership, and humor are the ingredients for survival in Ranger School. If it could be made light of, Jon and his buddies would poke fun at it and at themselves. One phenomenon familiar to all Ranger students is hallucinations. The lack of sleep and food tend to create weird things in the minds and eyes of the victims. Tim Rucker recalled that on one night patrol Jon thought he saw a Hershey bar hanging from a tree up ahead as they trudged through the swamps in Florida. Jon then quietly, so as not to attract the cadre's attention, whispered to Tim that he would take half and give his Ranger buddy the other.

In a letter Jon sent to me, now a Yearling at West Point, he provided his thoughts on the course once it was completed.

> *Ranger was rough, definitely, but we had a good group of guys and our class... did real well. 233 got the tab out of 272 that started, and only a few "grads" failed to get it. Right now the Arty [cadets that selected the Artillery branch] guys are arriving here for Airborne and Ranger. Winter Ranger [through the coldest winter months] is something I'm glad I missed.*

Jon passed three of his four patrols and graduated the course with honors.

## More School... and a Little Romance

Next stop—the Infantry Officer Basic Course (IOBC). A three-month academic course, IOBC teaches new infantry lieutenants the rudiments of training and then leading soldiers in combat. Covering every subject from small unit tactics to vehicle maintenance, the curriculum can be tedious and boring, for hardened students fresh from four years of college, though every subject is valuable for a rookie lieutenant.

In that same letter Jon sent me from the Officer Basic Course, he updated me on life in the Army to that point and asked many questions about friends and happenings at the academy. He admitted that he was writing the letter during one of his IOBC classes—maintenance—because it "helps to stay awake and helps to get something out of a pretty good class on a pretty dull subject."

Because Jon spent a concerted amount of personal time mentoring and training me in spiritual disciplines during his last year as a cadet, he was very interested in how I was surviving the rigors of academy life, particularly my continued spiritual development and faith. His tone was that of a spiritual big brother encouraging and exhorting a younger charge in his faith.

> *How is the situation up there now regarding your own spiritual life and that of the other guys (Ron, Percy, John, Barry B.)? I myself have had some pretty big ups and downs. The lack of a definite schedule and regular study and fellowship has been pretty hard for me to adjust to — but the Lord has as usual been faithful and I feel like I'm starting up out of a valley again. I'm continually impressed that we need Him as the center of our lives so much that not only do our consciences bother us, but we feel almost totally psychologically wrong when we're out of tune with Jesus.*

I have always marveled at the simple yet profound thoughts he conveyed in this letter.

His honest sharing of a struggle with a lack of consistent time to be with God, now that he was away from the academy routine, and the foundational principle of needing Christ at the center of his life, were a vivid demonstration of what he described as climbing out of the valley of complacency and inconsistency and back up the mountain to a more disciplined spiritual life in Christ.

Reality often comes into play after we've become spoiled in a generous spiritual environment where opportunities for growth and fellowship abound daily. The real world happens and forces us to reevaluate our stand and position in Christ. Jon's note was a powerful encouragement to me and made me aware of how susceptible we can become to apathy in the Christian walk. I didn't want to be apathetic or bored in my faith nor did I want to convey that attitude to anyone who might be watching my walk. I have never forgotten his words.

Though Jon's letter didn't allude to any bubbling romantic affair, he was shortly to make a momentous decision about his childhood sweetheart, Gail Morrison. She and Jon dated on and off from the time they were thirteen. Gail's recollections about her early relationship with Jon are particularly poignant.

> *We were both strong leaders in high school, very athletic, and <u>very</u> social beings. We were very active in our church choir and youth group. As a matter of fact, one Sunday night after choir practice, Jon and I were obviously trying to impress one another. We proceeded to get into our respective cars and initiated a side-by-side drag race down a local road. Alas, the sirens sounded and the lights lit up the sky and Jon and Gail were escorted to the police station. Thank goodness for small towns and solid records of good citizenship! We got off lightly but*

*were told in no uncertain terms that we were not acting as very good role models.*

She also provides a powerful rationale for her attraction to Jon. "Jon had a very gentle and quiet charisma. He had a photographic memory, got perfect SAT scores, was a talented athlete, an excellent thespian, had a glorious singing voice, a smashing sense of humor, was extremely handsome, had many devoted friends, and was very admired by his teachers," Gail recounted. "But because he was extremely self-confident yet humble, nobody was ever jealous of Jon nor did anybody ever have a negative thing to say about him. He was full of kindness."

Gail and Jon dated and corresponded throughout the first couple years of their respective four-year college experiences, Gail attending Syracuse University. Then something happened that often happens in a relationship supported mostly by letters—a friendly parting of ways—and perhaps a "valley" experience for both Jon and Gail. During their second year, Gail recalls, "Jon wasn't very nice, or particularly mature." She remembered thinking to herself, "Wow! He's actually human!"

Just before Thanksgiving in 1969, Jon invited Gail to join him at Fort Benning for the holiday. She did not hesitate, feeling "great elation."

At the end of their first day back together, Gail recalls a strange thing happening as the day drew to a close and they prepared to go to their separate accommodations for the evening. Jon looked at Gail with a serious gaze and said, "I just don't know about us."

Gail recalled years later, "His tone suggested to me that despite our friendship and love perhaps it was not the right relationship for him. Was I devastated!"

*Jon with Gail, his future wife, in 1965 just before Jon entered West Point*

It was not until the next day the couple discovered that Jon's choice of words and timing were ill-conceived. In fact, what he was trying to do, and eventually did, was convey to Gail that he wasn't sure that asking her to marry him *before* he left for Vietnam was the right and honorable thing to do. The harder right or the easier wrong? Certainly a life-changing decision faced them both. For Jon, to wait until his formal Army training and his year in Vietnam were over and then marry Gail perhaps was the right thing to do. They wouldn't be together much, if at all, during that time, and the uncertainty facing Jon as he prepared for a combat tour in Southeast Asia wasn't fair to thrust upon a young bride. Why not just let it go until this trying time in their lives was complete and their future much more certain and bright? That seemed the easiest choice to make.

Marrying her now—the truly harder choice—if she would agree, meant he would have to work very hard at learning the role of a new husband, while also learning the role of a combat leader during his short stint at Fort Carson, Colorado, and his further schooling before Vietnam—a lot to do in a short period of time. It would be one of the most challenging times of his young life. It seemed so selfish to put Gail through the hardship of a few months of marriage, only to have to say goodbye, knowing in all candor that they might never see one another again after that goodbye. Another issue that entered the picture was the contract that Gail had just signed with the school at which she was to teach. The thought of leaving her beloved children was agonizing.

That harder right was the only choice for Jon and for Gail. He asked Gail to marry him then and she accepted. As it turned out, the challenges of planning a wedding on such short notice and with such outside pressure from both their lives didn't faze these dreamers. Gail describes the feeling.

"Our wedding was traditional, joyous, and Christ-centered. We threw it together quite quickly yet felt none of that anxiety-provoking pressure. Every minute of planning was full of fun and anticipation."

*Out of the Valley*

*Jon and Gail's wedding day, February 7, 1970,
in Briar Cliff Manor, New York*

Their honeymoon was spent traveling across country en route to Fort Carson, Colorado, Jon's first Army assignment following his schools and training. As Jon knew, Carson was to be a brief indoctrination to the Army and the soldier's calling. It would allow him time with soldiers, some of whom had already served a tour or more in Vietnam. Others were, like him, getting ready to go. However, Fort Carson, like many other Vietnam-era Army installations, wasn't a shining example of the best and brightest the Army had to offer the nation. Serious problems of morale, discipline, and leadership were becoming evident, both in Vietnam and in stateside units. The growing lack of support from within the country for this war was powerfully evidenced by the shootings of students at Kent State University in Ohio, May 1970, only weeks before Jon and Gail reported for duty in Colorado.

It would just be that much harder to stay focused on the mission. It was now all about the war. Everything he did was to get himself ready to fight and destroy the enemy, while caring for his men. Was it worth it? Jon thought so.

Jon's life as a new leader of soldiers was very time-consuming and demanding, according to Gail. But commanding officers also knew that these young men would not have much more time with their wives before they shipped out to Vietnam, so some slack was built into their schedules. As all new military couples tend to do, Jon and Gail made friends quickly with their neighbors, some of whom were Jon's classmates and friends at West Point. One pastime the Shines enjoyed was riding a motorcycle into the foothills of the mountains surrounding the fort, and camping. They also attended a new church in the area and participated in Christian activities at the nearby retreat center and headquarters of the Officers' Christian Union (later Officers' Christian Fellowship). They grew deeper in love and deeper in their faith. It was a love and a faith that would be tested to its limits in the months ahead, with both venturing into some steep, dark valleys. But Gail remembers their relationship then as "in every way, very deep and very comfortable."

A good friend of Gail and Jon, Paul Pettijohn (former Executive Secretary of the Officers' Christian Fellowship), remembered some quality time with the two of them while they vacationed at the OCF's Colorado retreat center, Spring Canyon, located a couple of hours from Fort Carson. "I vividly remember going to their chalet to talk with them about getting ready to be apart and to prepare for his going into harm's way. Jon was very calm and he was spiritually ready. He was at peace with the task that was before him. The three of us talked about the role of the Word of God in our lives and ended up having a very meaningful prayer time together."

Paul also remembered a Scripture verse that Jon sent him in a letter from Vietnam in which Jon shared what became Jon and Gail's favorite verse—Romans, Chapter 8, verse 28: *And we know that all things work together for good to them that love God, to them who are the called according to His purpose* (KJV).

> AND WE KNOW THAT ALL THINGS WORK TOGETHER FOR GOOD TO THEM THAT LOVE GOD, TO THEM WHO ARE THE CALLED ACCORDING TO HIS PURPOSE.

"I remember as if it were yesterday," Paul recalled. "I felt the power and significance of what Jon was writing me. Second Lieutenant Jonathan Shine, U.S. Army, was saying in what proved to be his last letter to me, 'no matter what happens in Vietnam, I know it is going to work together for good.' What a powerful and profound application of God's

Word by a young officer who was going into his first round of combat!"

## Ethical Dilemma

Jon experienced some kind of serious dilemma during that time at Fort Carson. While we do not know exactly what he faced, it tore at him and he shared it with Gail.

Another valley... they seem to keep coming, without stop. Gail swore to Jon then that she would never reveal the dilemma's details to anyone. "It was very painful for him," Gail shared many years later. "He told me not to tell anyone. And it would be wrong for me to share the details now. It involves other people and somebody who was involved might read this and be hurt by it." Nearly four decades later, she has kept her promise.

We only know that it was extremely serious and caused Jon no small amount of anguish. It could have been an attempt by a superior officer to test Jon's personal and professional honor code. Perhaps there was pressure to "fudge" the monthly Unit Status Report that reflected the readiness status of the unit and could get a commander fired if it weren't up to par. It could have been a valued subordinate caught breaking a regulation or committing a serious breach of ethics and forcing his platoon leader to make a tough disciplinary call. It could have been a peer, committing an indiscretion and forcing Jon to either turn him in or turn the other way. Gail classified it as an "ethical dilemma that he had to confront."

Jon had been through some tough times before and knew it would not be an easy choice to make. When he finally made up his mind and chose a course of action, it evoked the following comment from Gail:

> *Despite his rank, he stood his ground and said he would not compromise his principles. He simply couldn't do that. There was no question about it... he would not be compromising. A higher ranked officer said, "I hope you keep your high principles, Lieutenant Shine. I doubt it, but I hope so."*

The Shines' time at Fort Carson was over before they felt they were settled as a new Army couple. Time was now nearer their point of separation than they would like to admit. The Army experience at Fort Carson was pleasant in some ways and frustrating in others. Who knows? Perhaps someday they would be reassigned to this beautiful Army installation and they could more fully enjoy all it had to offer.

For now, orders were in hand for the Jungle Operations Training Center in the Panama Canal Zone. While not Vietnam, it was the next best thing for replicating the rigors of maneuvering and fighting in the complex terrain such as they would likely find in Vietnam. It included triple canopy jungle, tall hills, and mountain ranges covered in dense vegetation, steamy days and steamier nights, monsoon rains much of the year, poisonous reptiles, leeches that sucked your blood until they were full, spiders as big as your hand,

and venomous ants as long as your index finger... all part of the tropical paradise that was Panama and that was soon to be experienced in South Vietnam.

Gail Shine remembered that last night together before Jon shipped off to Panama and, from there, to Vietnam.

> *The night prior to his going to Jungle School in Panama, we spent in New York City at a famous and very old hotel. We had an entertaining and romantic night "out on the town." However, back in the hotel I remember us holding each other, tears running down his face, as he said, "I hope this is where the Lord really wants me." He was a dedicated soldier. At the same time, he was a great thinker. It was expected that he go to Vietnam. He would have never done that blindly. He spoke often of political, moral, and ethical questions. When he made his decision to "sign up" for Vietnam I knew it was a very well-thought out decision.*

Jon left for Panama on August 15, 1970, and completed the two-week training scenario at the Army's Jungle Operations Course on August 29.

Next stop — South Vietnam.

# Chapter Three Roll-up

<u>Principles from Jon's Life</u>

- Mentoring brothers (or sisters) in Christ, via letter, is a perfectly acceptable and natural way to spend time, when you've already established a solid spiritual foundation in someone's life.

- God is always faithful and helps us adjust to new environments and challenging tasks that we will all certainly face.

- Choosing the harder right can be very painful, initially, but always results in discovering the ethical and moral truth… and the peace that God gives in tough times.

<u>"Harder Right" Choices Jon Made</u>

- Ethical dilemma at Fort Carson—he decided to maintain his integrity and not compromise his personal and spiritual values

- Asked Gail to marry him even though he was bound for combat in South Vietnam

- To seek Christ during his training and schooling, even though going with the flow of the freedom of a new lieutenant would have been much easier to do

## Putting and Keeping Christ at the Center

- Jon maintained a positive attitude and set a strong spiritual example while attending his training schools.

- Once you've worked with another in a spiritual mentoring/discipling role, you are drawn to that person in a powerful, spiritual way and desire God's best for him or her.

- Keeping Christ at the center of our lives through consistent time in God's Word, an active prayer life, and a focus on making disciples will get us through the valleys we encounter with strength, confidence, and ultimate victory.

## Discussion Questions

- What were some "valleys" at Fort Benning, GA (schooling) Jon either mentioned in his letter to Barry or were discovered later while at Fort Carson, CO, and how did he deal with them?

- What ethical dilemmas have you faced in the past or do you fear you might face in your future? What did you choose…or do you hope you'll choose?

- How did Jon stay connected with people he had mentored? How can you stay in touch with the spiritual development of those you choose to mentor?

- What was Jon and Gail's favorite Bible verse, and why did it give them both peace and strength as they faced separation?

- What is your favorite Bible verse…and why?

## Jon's Spiritual Legacy

- Classmate Tim Rucker

- Wife Gail

- Former mentee Barry Willey

- Senior ranking officer at Fort Carson, Colorado

# Chapter Four
# The Battle

Jon Shine arrived in South Vietnam in August of 1970. Ironically, combat began to decline in intensity in September 1970, as the United States began preparations for pulling its ground troops out of Vietnam. But the area known as the "Iron Triangle" became a hotbed of small unit actions as the North Vietnamese Regular Army prepared for a final offensive. Jon's company was placed under the operational control of the 11th Armored Cavalry Regiment (ACR) which was conducting frequent and aggressive reconnaissance patrols northeast of Saigon. (This 11th ACR was the same regiment which had supported Jon's brother, Al, in a major battle fought three years earlier.)

Nearly every day in the first two weeks of October, cavalry patrols came across signs of enemy activity—booby traps, bunkers, tunnels, equipment and food left hastily behind—hot trails. Little did this young Lieutenant know that a forthcoming encounter with a determined enemy force would be the ultimate "valley" experience of his life—into the valley of the shadow of death. Would his God be sufficient to help him out of it?

*A UH-1H helicopter carrying Jon Shine's platoon, landing in South Vietnam*

We get some valuable insight into Jon's character and strong sense of duty... and humor... from several of his comments to his new bride, Gail, taken from letters he wrote to her from Vietnam.

Here are some priceless excerpts from those letters:

> *We both know that my going is a matter of duty to our country, which has given us so much. This sacrifice on our parts is certainly a small one for the affluence and freedom we have enjoyed.*
>
> *You can't imagine how much I want to keep each of my men safe. I think if we killed one hundred enemy soldiers and lost one man, I would*

*consider it a defeat. Maybe that's not too military a way to feel.*

[Author's note: This is a real-life "valley" that all combat leaders face and a tough one to resolve internally… do I go all-out and aggressively seek out a fight… or do I work harder to create an effective defensive posture so we all come back safely home?]

*Guess what? I discovered Sunday that 1LT Ralph Garaway of 1st Platoon, C Company is a Christian! Just shows how much each of us is witnessing when we can be together for three weeks and not know that the other is saved! It shows me again that God answers prayer.*

*I think the Lord must be smiling down on us. We've found about 20 booby traps in the last three days and no one has been hurt.*

*As far as danger goes, I praise God that we can know that we are each in His hands and that nothing will happen to either of us that is not in His plan. We don't have to worry about "odds" or "chances" or "accidents." I will be just as much in His care in Vietnam as you are in New York. I really do agree with you that this year should be a time of spiritual growth for both of us. The true impact of Romans 8:28 is just beginning to hit me. What a wonderful promise!*

Jon had been in combat operations for only a few weeks when he wrote to Al, his older brother, about a desire he felt to look out for the men in his platoon and keep them from becoming casualties as politicians back home sorted out U.S. policy toward Vietnam, the President prepared to start bringing troops home, and many back home protested the war in America's streets. He knew that wasn't how things were supposed to be, but he was being honest with Al.

*Jon Shine in Vietnam*

When Al received Jon's letter he immediately penned a reply that mildly scolded Jon and lovingly but directly charged him with the firm responsibility of taking the fight aggressively to the enemy.

To close with and destroy the enemy was the infantryman's mission... Jon's mission. No matter what was going on in the halls of Congress or the streets of America regarding the war, combat leaders were to remain focused on the mission at hand—find the enemy, fix him, and fight him.

Jon never received Al's letter. But he really didn't need it. He knew in his heart what he had to do.

## Into Action

On the 15th of October, 1970, Jon's platoon was working "split-platoon" operations whereby a roughly 30-man infantry platoon split and covered more terrain looking for enemy soldiers. They would then re-group as needed. Jon was leading one half and Sergeant Greg Yahn the other.

Sergeant Joe Christopher, one of Jon's non-commissioned officers, left his four-man reconnaissance element at a concealed location, only yards from eight North Vietnamese Army regulars, their radio blaring and the soldiers oblivious to the U.S. infantrymen nearby. M60 machine gunner Carl Nichols got specific instructions from Christopher not to fire unless it was absolutely necessary. Christopher then rejoined his main element, led by Sergeant Yahn, about 30 to 40 meters away. At the same time, Lieutenant Shine, leading the Third Platoon in Charlie Company, 4th Battalion, 9th Infantry Regiment (the Manchus), 25th Infantry Division, linked up with Sgt. Yahn when he heard about the enemy sighting.

The men quickly discussed in low voices how they should handle the enemy force. A typical tactic for this kind of contact was to pull back a safe distance and call in artillery, helicopter gunships, and jets to unload their ordnance on the unsuspecting enemy. Jon Shine's small force could certainly count on the help of their higher headquarters' arsenal to cover their actions. Another possibility was a frontal assault, achieving shock action and hopefully a quick, decisive victory, but a very risky venture with high probability of casualties. A third course of action involved an aggressive attack on their flank, thereby gaining some measure of surprise while less likely to result in serious casualties.

RAT-TAT-TAT-TAT-TAT… the unmistakable sound of the infantryman's best friend in battle—the M60 machine gun! Something must have happened to cause Nichols to fire off a burst of thirty rounds. No more discussions of options. It was now time for action. Jon called out to his men as he literally lurched toward the enemy's flank, leading the rest of the platoon in this gutsy move.

In pursuing the smaller enemy unit, however, they soon discovered they had run into that unit's larger force, a huge enemy bunker complex with what was later determined to be about 100 NVA regulars and two 30-caliber machine guns trained on them. The action quickly turned into a larger firefight in which the 3rd squadron of the 11th ACR eventually became decisively engaged.

The ground trembled and opened in wide gaping holes as North Vietnamese Army mortar rounds landed near the men of Jon's platoon. The deadly projectiles, lobbed with precision accuracy from perhaps one terrain feature away, were joined by rocket-propelled grenades, arcing into their oblong piece of ground, exploding into hundreds of molten-hot fragments. Man-sized chunks of mud rose from the earth-like geysers each time a round landed. NVA 30-mm machine guns and AK-47 assault rifles joined the cacophony that became a roar in the ears of Jon and his soldiers.

Sgt. Greg Yahn shared some memories of that battle:

> *We had just decided on a compromise of Lieutenant Shine taking a flanking position, when the firing started. The lieutenant sprang into action with Sergeant Roberts following. They led their platoon on a flanking position from where our main group was set. The jungle was thick, from anywhere just off the path, and made visibility of the enemy past 10 meters impossible to detect without observing muzzle flash. As he rallied his troops to move to our right, they maybe made about fifty yards progress, when he was cut down by machine gun fire.*

Rob "Doc" Jackson, Jon's platoon medic, unhesitatingly moved to the front of the column when word came to him that casualties had been sustained. He began intently working on Sergeant Joe Roberts, one of Jon's squad leaders, as Jon lay five feet away. Both men were seriously wounded from the initial enemy fire. Roberts had taken two bullets in the chest and Doc feverishly tried to stop the flow of blood, deal with the "sucking chest wound" that comes from penetrations of the lungs, and treat for shock.

Oblivious to the enemy fire all around him, Doc knew his lieutenant was either wounded or dead nearby. At this point, Jackson's remembrances are powerful:

> *As I was trying to bandage his wounds and assess what was appropriate to do next, I heard a voice just a few feet to Robert's left. Realize all this was happening in intermittent hails of fire like torrents, but somehow I could hear Jon very clearly. It was for me a very special moment, it was holy, and I realized it even then. I had just been saying the name of Jesus out loud, over and over as I worked on Roberts, and I heard Jon say, "Doc, I've been hit in the head but I'm okay. Just throw me some bandages and I'll stop the bleeding until you finish with Roberts and get back." That's very close to verbatim.*
>
> *Over the years I've told the story many times and I always include how remarkably composed he was. It wasn't until I met you, Barry, that I understood how he could be so calm, so secure. It was, of course, Christ in him the hope of glory. What a privilege to be there with him even for those few moments. I threw him the bandages and while I was dragging Roberts back he was killed and I never saw him again. Lewis Lesnikowski and Gene Hess both got up to him, or close, but he was gone home and in the Savior's presence, as we labored on.*

Platoon member Ted Hooker affirmed Lieutenant Jon Shine's desire to help Sergeant Roberts: "I remember the lieutenant pleading for someone to help out Joe... definitely saying 'Get Joe. I'm OK. He's hit in the chest. Get him first.'" Ted also remembered possibly seeing Jon with a .45 caliber pistol about to engage or possibly engaging the enemy at some point in those confusing moments after Sergeant Roberts and Jon were hit.

From Jon's life story, we now know of his propensity to sacrificial living and action when others are involved. He constantly chose the harder, more risky, more dangerous "right" rather than the much easier "wrong" when confronted with such dilemmas. We have seen that he moved—yes, bolted—into action, without any hesitation, toward the enemy that was engaging and threatening his small band of soldiers, 75 yards from his platoon position. He could have tried to call in artillery and jets to bomb the enemy force, but his troops were too close and that would take too long and even now they were engaged in a life-or-death struggle at close quarters.

The citation to Jon's posthumous Silver Star for gallantry in action described his actions: "During the initial contact, Lieutenant Shine was seriously wounded. Despite his wounds, Lt. Shine immediately began placing suppressive fire on the enemy positions, thus allowing his men to move to cover." His words to "Doc" Jackson seem clearly intended to keep him and the other platoon members focused on Joe Roberts for the few moments that he engaged the enemy.

Jon, thinking only about his men and acting on their behalf, perished when the enemy returned his fire.

When word of the fight reached back to Cu Chi base camp, an incredible thing happened. The battalion scout platoon was just back from an operation for rest and recuperation. Their leader was Lt. Bill Yonushonis, a close friend of Jon and a West Point classmate, and his soldiers knew of that strong bond and Jon's reputation in the battalion.

Without orders, they put on their combat gear, drew ammo, and stood by to go in and retrieve Jon's body. The battalion commander himself had to order them to stand down. What a powerful testimony was this act of brave men, whose love and respect for a fellow soldier led them willingly to this point of collective sacrifice.

## A Gray Day at West Point

The rain drizzled down and soaked my heavy full dress cadet uniform jacket. The rain dripped from my cap bill down in front of my eyes. The rain mixed with tears as we gathered for Jon's funeral on that dreary day in October of 1970. I was so very proud to have known Jon and been a part of his life and he mine during my first year at West Point. I pondered the loss we had experienced. I hadn't grasped its meaning yet or God's eternal purpose in it all. I was just sad, angry, and uncertain about everything.

Why had God taken him when he was so young and had so much to offer? I remember saluting the coffin carrying my friend as it descended the steps of the Old Cadet Chapel, in the midst of the cemetery at West Point.

We marched along slowly as we followed the hearse toward the burial site. Family, friends, and fellow cadets were all huddled under an awning erected to cover the burial site and protect from the rain. Many others who could not fit under the awning stood around it… silent… heads bowed or following the funeral marchers as they approached the grave.

*First Lieutenant Jon Shine's funeral procession at the West Point cemetery on a rainy day in October 1970*

At some point during this melancholy time frame, I received a special insight, from the Holy Spirit I am very sure, that gave me a peace that I had not really known before. It was an assurance that Jon had accomplished all that God wanted him to accomplish on this earth and was now with Him in heaven. It was also a strong sense that I was to carry on Jon's spiritual legacy... starting right then. And the Lord would walk beside me all the way. That gray, somber day turned out to be a bright day of inspiration and motivation for me. I took a deep breath, sighed, and set my heart on serving Him.

*Air Force fighter pilot Tony Shine, Jon's oldest brother, escorted Jon's body back from Vietnam and presented a flag to Helen Shine, their mother, with Jon's other brother, Al, to Helen's right and Al's wife, Sandra to her left.*

*Al, Jon's older brother, presents the flag to Jon's widow, Gail, with Mrs. Helen Shine, Jon and Al's mother, looking on. The author, Barry Willey, a cadet at the time of the funeral, looks on from far left.*

## Heroes All

In May of 2002, I had the honor and privilege to join seven men from Jon's platoon in Vietnam and share in their reminiscences of their leader and his final battle. Greg Yahn, Gene Hess, Joe Christopher, Jesse 'Sal' Salcedo, Rob Jackson, Steve Harlan, and Ted Hooker were plain-talking heroes who gave their all, and after Vietnam went about their lives as solid citizens, not asking anything from their country that asked so much of them. Rob Jackson faced almost certain death in the face of horrendous enemy fire but focused on taking care of the wounded at the scene of the fight. Gene Hess courageously attempted to rescue Jon after Jon was hit the second time and Gene was seriously wounded in the attempt.

Greg Yahn bravely went in and pulled Gene out. Joe Christopher was wounded as he followed his leaders to the sound of the guns. As he was being lifted out of the jungle in a litter, the helicopter had to cut him away because it was hit and going down. In addition to his wound, he suffered multiple injuries and life-long physical and emotional challenges as a result of that 60-foot fall. Sal Salcedo, Steve Harlan, and Ted Hooker all heroically engaged the enemy force and helped their buddies throughout this entire ordeal. The enemy fire was so intense that no one was able retrieve Jon's body, even after the reinforcements from the 11th Cav arrived in the area, until the next morning. At that time, the enemy force had vacated the area, probably by boat from the rear of their position which butted against the Saigon River. All of these men gave all they had that day and deserve our nation's most profound gratitude.

Steve Harlan told me, "Good men died that day."

Greg Yahn wrote the following poignant words about this "book effort" to honor Jon and his legacy: "Lt. Shine was absolutely courageous in his assault on that day, as was Sergeant Roberts. I wish this book could be written from the first person, because we have all lost a comrade soul. God bless you and your family. God has directed your hand in all of this, pulled all of us veterans together, helped to give us rest in heart and mind, and given us all a reason to remember each other. For this I am forever grateful."

God also directed the hand of a young Army Major, Bill Tozer, who was the Casualty Notification Officer for the Shine family. Bill didn't know going into this tough mission the grace and love he would experience from the entire Shine family, but discovered it nonetheless. His amazing and poignant story is included in this book in Appendix 2, Testimonials.

*Jon Shine and "Country Joe" Roberts' names are on the Vietnam Memorial Wall, panel 6 west, line 2... together... just as they fell in battle*

# Chapter Four Roll-up

## Principles from Jon's Life

- The mentally, physically, emotionally, and spiritually disciplined life can accomplish amazing things, for mankind... and for Almighty God.
- Servant-leadership is real and is demonstrated in those who are committed to a Christ-centered life.
- Making the harder right choice is always harder when the chips are down and the stakes are high. But doing the right thing pays off in changed lives.

## "Harder Right" Choices Jon Made

- He chose to attack the enemy position with his platoon, despite the prevailing thought among the U.S. forces in South Vietnam at that time that the war was winding down quickly and the best action was to take no action... keep your men safe and bring them all home. Don't be aggressive with the enemy. The war is almost over.

- After being wounded in the head from the enemy's first volley of fire, Jon directed his men to take care of another wounded soldier and get him evacuated... then they could come back for him. In that brief moment, while they tended to the sergeant, Jon opened fire on the enemy position to cover his men's withdrawal and was killed when the enemy returned his fire.

## Putting and Keeping Christ at the Center

- Jon exhibited a Christ-like attitude in South Vietnam when he demonstrated the true meaning of John 15:13 (NIV), "Greater love has no one than this, to lay down one's life for one's friends."
- Rob "Doc" Jackson, Jon's platoon medic, remarked for years after that battle and hearing Jon's final words on earth, that Jon had such a peace in his voice and demeanor when he asked Doc for a bandage and directed the men to take Sergeant Roberts off the battlefield. Doc told the author that he knew the reason for the peace in Jon Shine's life… "It was Christ in him, the hope of glory."

- Jon's favorite Bible verse was Romans 8:28 (KJV, *And we know that all things work together for good to them that love God, to them who are the called according to his purpose.* He surely knew as he lay wounded and in desperate straits on that battlefield that God was, indeed, going to work this out for good… and He did.

> AND WE KNOW THAT ALL THINGS WORK TOGETHER FOR GOOD TO THEM THAT LOVE GOD, TO THEM WHO ARE THE CALLED ACCORDING TO HIS PURPOSE.

## Discussion Questions

- What valleys did Jon find himself facing in this chapter, and how did he climb out?

- What could motivate someone to sacrifice his or her life for someone else?

- Do you have that peace that passes all understanding that Jon apparently had as he lay wounded on the battlefield? Why did the medic who treated Jon call that peace... "the hope of glory" for believers? How do we receive that peace in our lives? (The verse to which this alludes is Colossians 1:27.)

- What kind of insight and inspiration did Barry receive at Jon's funeral? Have you ever gotten an insight or inspiration from God at any time in your life? Describe that experience.

- How does God use a terrible situation in our lives... for good? How does He turn bad into good?

- How do we know if we have been "called according to His purpose"?

## Jon's Spiritual Legacy

- Rob "Doc" Jackson
- Bill Yonushonis
- Greg Yahn
- Steve Harlan
- Joe Christopher
- Jesse "Sal" Salcedo
- Ted Hooker
- Gene Hess

# Chapter Five
# An Eternal Legacy

In the months following Jonathan Cameron Shine's death, his courage, compassion, love for the Lord, and commitment to mentoring God's Word provided an invaluable and uplifting continuum for me in military and spiritual life, even as I, along with so many others, mourned Jon's physical absence. We all drew strength and hope from inspiring memories... from the eternal legacy of which Jon clearly was, and is, such a part.

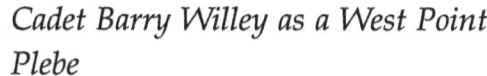

*Cadet Barry Willey as a West Point Plebe*

For me, my sophomore (yearling) and junior (cow) years at the academy were exciting years of physical, professional, and spiritual growth. My desire to serve the Lord and mature in my faith increased daily. The importance of Bible study, a consistent prayer life, fellowship with other believers, and sharing my faith with those around me were anchor points in my life at West Point.

My spiritual life was developing from the solid foundation that my wonderful parents, Carroll and Mary Willey, had built in me from my very early days. As a result of this spiritual growth, my desire to spiritually mentor other men, as Jon Shine had done with me, increased with my faith. Finally, during my senior (firstie) year, I asked two younger men in my battalion—Greg Schumacher and Jim Blackwell—if they wanted to join me in a year of intensive spiritual growth together. They both eagerly agreed.

*Fort Shine at Spring Canyon in Colorado*

My desire was to share the disciplines of the Christian faith that I was taught from my parents, and that I had learned from and seen so clearly in Jon and the Christian faculty officers who had "adopted" me. I would now pass them on to Greg and Jim, so they in turn could pass them on to other faithful believers. I was convicted and convinced of the truth in Scripture that the Apostle Paul taught to his young friend, Timothy (2 Timothy 2:2). It was time to get serious about a spiritual life-calling that would motivate me and pervade my conscience in all I did for the next 29 years as an Army officer and three years as a corporate consultant. It would be the compass that guided me and gave me a purpose beyond all other purposes in my life. The Lord told us what He wants us to be doing while we spend our time on earth. He said, "therefore go and make disciples of all nations."(Matthew 28:19) I realized I needed to be about His work.

Jim and Greg were, and still are, wonderful friends and brothers in Christ. They matured in their love for the Lord and His service and have served Him faithfully all their lives.

A couple weeks before I graduated, I had the privilege to fly to the OCF western conference center—Spring Canyon— just outside Buena Vista, Colorado, to participate in a very special ceremony—the dedication of a newly constructed lodge named Fort Shine in Jon's honor.

In the years since, literally thousands of Christian retreaters have rested in that lodge and have been touched by the strong spiritual legacy Jon Shine left behind.

*Then-cadet Jim Hougnon spoke at the Fort Shine dedication at Spring Canyon in 1972.*

Following graduation from West Point in June of 1972, I had the distinct joy to travel with several Christian buddies and the founder of OCF in the U.S., Cleo "Buck" Buxton, throughout Europe and the Holy land for three of my vacation weeks before reporting to my first schools and my initial assignment. That time was a powerful period of growth as a believer, learning at the feet of a man like Buck and just being together with close brothers in Christ.

After completing the Army's Airborne and Ranger schools and my branch schooling—Infantry Officer Basic—I reported as a second lieutenant to the storied 82nd Airborne Division, the quick reaction force for the United States. I was humbled yet proud. Being a hard-charging platoon leader in an infantry battalion in 1973 was a challenge. The soldiers were tough and motivated but also affected by the culture of the late '60s and early '70s within our nation. Iron discipline was needed in a unit like the 82nd Airborne, and leaders were expected to set the example, train, and lead their men with integrity, wisdom, and solid values.

In addition to the end of the Vietnam conflict, another world event during 1973 was the Yom Kippur War involving Israel. I can remember being alerted one day to report to our unit area and prepare my platoon for possible deployment to Israel, along with the rest of the Division, to support our ally in that war. Emotions were high... excitement, wonder, and yes, fear. Some soldiers deserted... went AWOL. But most were pumped and ready to go. As it turned out, we were ready to load the airplanes for the flight to Israel—when it was called off. Nevertheless, we all felt what it was like to be physically and emotionally prepared for combat. We also became sensitized to the spiritual side of our makeup, a side often ignored by leaders who are reticent to venture into that uncomfortable or embarrassing arena. Or they simply aren't sure about their own faith and can't see themselves trying to coach their men in spiritual things.

The Chaplain Corps, of course, exists to take up where an individual leader feels inadequate or unable to help his men spiritually. I didn't feel that I could do a better job than my unit chaplain to prepare my soldier's spiritually. I did feel, however, an obligation to share my own faith story and ensure they knew that if they ever needed encouragement in that area of life, they could call on me without hesitation. I was also still very cognizant of Jon Shine's impact on my life. He was killed in action only three years earlier. My memory of his brief but inspired life and my own changed outlook on life—feeling a strong need to work with men as a spiritual mentor—convinced me that I needed to share my Christian testimony with my platoon of about forty soldiers.

I called them together one day after training, shortly after the Israel alert, and sat them on the steps to our barracks. I spent about ten minutes sharing my personal philosophy of life as a Christian and was compelled to talk to them about my relationship with Jon Shine, his powerful touch on my life, and his heroic death in action during the Vietnam conflict. Some squirmed; others looked away...but all listened. Most seemed appreciative of my willingness to share something many of them had never heard before—nor expected to hear. I wasn't quite sure what to expect following this session but hoped a few of them were sensitized to the Christian lifestyle and the hope and assurance it provided.

> *This approach to sharing one's Christian faith in the unit area, in uniform, may not be the recommended method in today's very sensitive "separation-of-church-and-state" culture. There will always be opportunities — after duty, not-in-uniform, in non-duty locales — when one can share his or her faith more openly with less risk. If asked by a platoon member or peer or senior, it is certainly appropriate to share one's faith directly and in accordance with local military protocol.*

I remember Jon and others in my early days as a believer at West Point asking, "How best should we be spending our time in this life?" Their own answer was powerfully simple — "by focusing on the only two things that really last…that really have eternal value…the Word of God and the souls of men."

If you think through that statement, it makes a lot of sense. Everything else, no matter what it is — fame, money, stocks, bonds, Certificates of Deposit, IRAs, 401(k)s, perfect bodies, great jobs, PhDs, great art, or literature — it's nice to have while here, but it is all meaningless when life is over. You can't take any of it with you! What lasts are the words of our heavenly Father, found in the Holy Scriptures, and the souls of men and women… that which passes on into some kind of eternal state after death on this earth.

The session with my platoon was on a Friday afternoon. That Sunday night I was at my apartment in Fayetteville, North Carolina, getting my equipment ready for a field exercise the next day. It was probably about 9 or 10 p.m. The phone rang and I thought it a bit unusual to be getting a call at that hour. I answered it and the voice on the other end said, "Lieutenant Willey, sir, this is Specialist Fred." He was a machine-gunner in my rifle platoon and not one of the stellar performers. In fact, there were rumors going around the platoon that Fred was into drugs, both using and pushing, and we were simply waiting for an opportunity to catch him in the act and remove his bad influence from the other young troops. Now, I figured, was our chance. I suspected he was calling from the jail downtown and needed me to bail him out.

"Sir," the young soldier continued, "I went to South Carolina this weekend with a friend from our platoon and met someone while I was there. I met Jesus Christ while attending a Billy Graham film and gave my life to Him. I was wondering if you would help me learn more about Him since you talked to us about Him on Friday."

Needless to say, God had worked a miracle in young Fred's life through the power of His Spirit and a power-packed Billy Graham film. And he knew about my faith because of my taking the time to share my testimony. I assured Specialist Fred that I would gladly help him learn more about the Lord.

I recalled the early days of my first year at the Military Academy and Jon Shine—my senior by three years—taking some risk by teaching me about faithfully walking with the Lord and being a disciple. I wanted to follow that example and be a spiritual big brother and mentor to this soldier. Because he was a member of my platoon and therefore in my chain of command, this situation presented me with a potential challenge—possible charges of favoritism or fraternization with an enlisted man. I very discreetly met with him after normal duty hours for several weeks and helped him learn a few spiritual ropes before my job changed within the battalion. Jon Shine's influence was taking hold and motivating me to follow in his footsteps...and as Jon would have observed, more importantly, to follow in the Lord's footsteps.

I saw in that experience, orchestrated by the Lord, that God had a plan and purpose for me as I progressed in my military life. He wanted me to be the best Army officer I could possibly be, while seeking His kingdom first and working with men desirous of being disciples. Those goals were fully integrated, not mutually exclusive. He also had a plan for me to marry the right person—a wonderful, godly, faithful woman and the perfect mate, Barb—and establish a family. I did that in June of 1974, and together we started a family and have tried to serve the Lord since that day. Rachael and Jonathan came along in 1975 and 1977, respectively, and have been wonderful blessings to us. (Jonathan, of course, was named after Jonathan Shine. Praise God that both of my children's testimonies are powerful tributes to Jon's inspiration—directly in my life and indirectly in theirs.)

## Two Valley Experiences

The main focus now is how the Lord has worked through faithful men and women to build His kingdom upon the eternal legacy that Jon Shine left us. At this point, however, it is important to share how the Lord has lifted me out of a couple of precarious "valley" situations. While this story is not about Barry Willey, my purpose here briefly is to brag on the Holy Spirit and His moving in my life when the outcome looked grim and uncertain. I also want to share how Jon Shine's memory, and the assurance I have of knowing how God worked in his life, was an ever-present comfort to me.

As a U.S. Central command staff officer, while serving aboard a U.S. Navy combat ship in the Persian Gulf in 1987 (during Operation Earnest Will, when U.S. flags flew on Kuwaiti oil tankers to allow us to protect them legally), a violent storm blew up one night. I stood on the ship's bridge wing, a narrow walkway with a small railing, just outside of the place where the ship is steered—the bridge. The ship rocked and rolled as large waves blew over the side. I couldn't get to a place of safety before I was overcome by sea-sickness and had sunk to my knees out on that narrow walkway. I was seconds from being washed over the side and unable to move, due to nausea and weakness. My only course of action was to look skyward and utter a very simple three-word prayer... "Lord, I'm yours." Within five seconds, the nausea was gone! And as quickly, a wave of strength flowed from my toes, through my legs, and up my body. As it did, I rose to my feet, released the railing I had a death grip on moments before, and walked down through the ship's bridge, down the ladder, and to my bunk. I strapped myself in, as the ship was still violently rolling. I was asleep in about a minute.

Another time, during Operation Desert Storm, I was leading the first group of soldiers from the 24th Infantry Division out of Iraq and back to the United States. We were the first heavy unit to deploy on Operation Desert Shield about seven months earlier and the President of the United States chose us to return first. Getting out of there was pretty dicey, though.

We were sent to a destroyed Iraqi airfield and were to await U.S. transport aircraft to take us back home. The airfield was littered with unexploded bomblets from our bombing runs, and a U.S. Air Force Explosive Ordnance Disposal (EOD) Unit was trying to clear the airfield of the dangerous munitions. While reconnoitering the airfield in my Hummer vehicle, we had a flat tire. My driver stayed to repair the tire while I bummed a ride from one of the EOD airmen. As he drove down the runway with me in the passenger seat, he suddenly swerved to the left and hunkered down instinctively.

A few seconds later, I asked, "What was that all about?" He explained with some emotion that we had just driven by a magnetic influence mine and it should have exploded… but it didn't. We went about our mission but I have always firmly believed that mine failed to explode because God had other plans.

## Faithful Generations Continue

Another "generation" of believers empowered by Jon Shine's example includes Timothy Mallard, a young man I met with when stationed in Panama. Timothy is now an Army Major and a Chaplain who deployed to Iraq with the 101st Airborne Division (Air Assault) during the original Battle for Baghdad in 2003 and is on fire for the Lord and His work today.

He recently shared his personal Christian testimony with me, and here is how it started:

> *My story as a Christian goes back to several formative experiences, not the least of which was a Bible study and discipleship program I experienced in Panama with Barry Willey when I was a teenager. Many years after that, however, I dedicated myself to full-time Christian service, followed a call to the ministry, and another call to the Army chaplaincy.*

While a Chaplain at Fort Benning, Georgia, for a mechanized infantry battalion, Timothy also led a ministry at a small chapel on the base. He was soon reassigned to Europe and found himself in a Germany-based artillery unit bound for Bosnia.

He was encouraged by some of his soldiers to start a Promise Keepers Bible study and found himself ministering to a diverse group of men from all races, religious affiliations, and backgrounds, all desirous of learning about and walking close to Christ. They started calling themselves the "Men of Integrity." Timothy eventually led this group of men on a spiritual journey to Washington, D.C., during the Promise Keepers' "Stand in the Gap" rally in October of 1997. It was a logistically challenging time to fund, transport, feed, and house a group of thirteen soldiers, but through some amazing answers to their prayers, it all worked out and they experienced a life-changing time together.

Timothy concluded his testimony this way:

> *We made it back to Germany and, returning to our community, began to share with others our adventure in Christ, including those brothers of ours who had remained behind to accomplish other missions. I knew then, however, that our fellowship — not just those who made the journey, but all fifty of our men — had achieved that which God had purposed for us to accomplish. I knew that it was time for us to begin going out from that place to carry our faith to others.*
>
> *Not eight months later over half our group would be gone to other parts of the world. We have eight men preparing for or serving in full-time ministry, two who have gone to college to return to the Army as officers, and many who are in new*

*places of service in the local church as musicians, deacons, lay leaders, or teachers. As well, there is no telling what impact on the world the sons and daughters of these men will have in the future.*

One of those Men of Integrity we do know about is Sergeant First Class John K. When I contacted him—another "generation" of faithful believers following Jon Shine's example—in September of 2001, he wrote this response:

*I first met Chaplain Mallard in Bosnia in a Tactical Operations Center (TOC) in 1996. We had a nightly briefing at 1900 hours and it was his first night there. I remember him standing in front of everybody in the TOC and giving us the Scripture of John 15:5, "I am the vine; you are the branches. If a man remains in me and I in him, he will bear much fruit; apart from me you can do nothing." Now I remember this because at that time I was searching to find out who this God was. I gave my life to Christ and recognized Jesus as my Lord and Savior 20 July 1996. That Scripture has stayed with me until this day.*

*There was something different about this man, different from other chaplains I had met previously in the battalion. At the time I didn't know what it was, but later realized that he had a light that shined. There was something in this man's life that I wanted, too. He was compassionate and very friendly... always had time for others... and what was best is that he brought people together. We had services in Bosnia and people came to hear what this man of God had to say. Church was never that full before, but now they were coming. They must have seen what I saw and that was the realness for Jesus Christ.*

I knew at the time I contacted John that he was heading up a group of men at Fort Hood who were preparing to host a Central-Texas-wide Christian men's conference. They had planned it for months and briefed the concept to the Garrison Commander, receiving his approval to proceed. When I reestablished contact in April of 2002, John and his group had just completed the conference, whose guest speakers included author Stu Weber and pro-football great Hershel Walker.

John's note to me said:

> *...we just had a wonderful, blessed time in the Lord...What I found really great about the whole conference was the prayer leading up to the conference. We asked the Lord if just one came and gave his life, all would be worth it. Well, on Friday night the altar was open and we had about 25, including a nine-year-old who was moved and came by himself.*
>
> *On Saturday before the conference started, we had a soldier who was running the track at 0530 and the praise team had just finished setting up and started playing. This individual started asking the*

*conference staff what was going on. We explained to him what we were doing and one thing led to another and BAM!!...out of the clear blue he wanted Jesus in his life. We called all the men around who were there and prayed with him. I thank God every day that He can use men all around us for His glory."*

Timothy and John continue today as faithful servants for Christ in the Army and their respective communities. They also represent Jon Shine's powerful spiritual legacy that is living and actively building God's kingdom here on earth.

Here's one more amazing note on Timothy Mallard. For the past several years, he has ministered to his community as an Army chaplain in Germany. In addition to his pastoral duties and responsibilities, he and his wife, Sharon, have led an Officers' Christian Fellowship Bible study and discipleship group in the homes of members of their unit. At the initiation of his small group several years ago, Timothy shared this with his group members:

*I have contacted each of you over the past month regarding your interest in beginning an Officers' Christian Fellowship Bible Study. We now can come to the start time and Sharon and I invite each of you and your spouses to our home tomorrow night at 1830 to begin a new work in the Lord. As I see it, our group will have five critical purposes:*

*1. Strengthen our relationships individually and as couples with Jesus Christ, our Lord and Savior;*

*2. Be obedient to His command to teach and make disciples throughout the world;*

*3. Study God's Word in an atmosphere of collegial support to strengthen our knowledge of the Bible;*

*4. Pray diligently for each other and His Church and support one another in our daily lives; and*

*5. Prepare for whatever follow-on spiritual task that God calls us to in our Army careers.*

*Brothers, this is no meager spiritual effort. God can and will do great things through us. You'll notice that I have cc'd above COL Barry Willey, US Army, Retired. Barry is the OCF Director at West Point, but more importantly, Barry is my friend and mentor, a man who took me aside for two years and discipled me in the Word. My ministry now in the Chaplain's Corps, and more importantly my vital relationship with God through Christ in the power of His Holy Spirit, is the direct result of Barry's effort. I have prayed and will continue to pray, that God will so bless us that years from now, our spiritual ancestors will look back on this time as formative in their walk with Him.*

I hesitated to include some of the comments and quotes from friends in this book because they talk about me. I am humbled, and yes, embarrassed, by accolades that lift me up in any way. I only include them because they point back to Jon Shine's influence on many generations of Christians while he was here, and, ultimately, on the power of the risen Christ in Jon's life to do what he was able to do.

God began this ministry of multiplication through generations of faithful men years ago, using Jon Shine as a powerful influence on many others along the way. We don't always know how or why He works the way He does, but we can be assured that He is in control. He took Jon from this life, at age twenty-three, to be with Him. Jon's tragic death was certainly a horrible loss to his young wife, his family and friends, and to the Army and his country for which he had so much potential. We may never know what Jon could have accomplished in an earthly sense. But we do know what he already has accomplished in a spiritual sense. Jonathan Cameron Shine serves as a life-altering inspiration and motivation to live for Jesus Christ and serve others selflessly and sacrificially.

The time Jon Shine invested in me—studying God's Word, praying together, meeting in fellowship with others at the Academy, and going to spiritual conferences on weekends—made profound inroads in my life and set me on a course of personal spiritual discipline and training of others that continues to this day, back at the place where it all started—West Point. And as we have seen, Jon's desire to meet with, teach, and spiritually mentor me wasn't happenstance or whim. It was the result of God inspiring him to participate in a discipleship program with Paul Stanley and his subsequent heart's desire to share and invest in a few of us who wanted to grow in our faith.

*Barry and Barb Willey joined Officers' Christian Fellowship field staff at West Point in January of 2005*

## OCF at West Point

In 1975, only three years after my graduation from the Academy, Barb and I found ourselves, unexpectedly soon, in an assignment at West Point—I was the Aide-de-Camp to the Commandant of Cadets. When I wasn't "aiding" and Barb was not occupied with our first child, Rachael, we worked with Dr. John and Judy George, the staff representatives for OCF at West Point. Our ministry then was called the Timothy Club (after the 2 Timothy 2:2 principle of disciple multiplication).

It was exciting to work with the First Cadet Regiment— my old regiment—where we saw many young cadets deepen their faith in Christ, following in Jon Shine's footsteps. John and Judy were a blessing to Barb and me and to all the cadets with whom we served.

It was actually Judy's insistence that Barb pray about, and consider putting our names in the hat for, the position of OCF staff representatives at West Point when it came open in 2005. In fact, we did and were wonderfully called by God to go back to West Point to serve there for OCF in January 2005.

While we were there, we got more exciting news and clear evidence of God's hand in building upon Jon's spiritual offspring. Chaplain (MAJ) Timothy Mallard, previously mentioned, who has had such a powerful impact on many lives within the Army as a chaplain, was assigned to West Point to be the Senior Cadet Chaplain. In that capacity, he had the opportunity and joy of continuing on, with Barb and me, this amazing work of God — building eternally into the lives of the most capable of our nation's future leaders.

OCF is an international organization that seeks to teach and train cadets, staff, and faculty to exercise biblical leadership to help raise up a godly military as they enter the Army and seek to serve Jesus Christ. Many cadets love the Lord when they arrive at West Point, and we are able to help better spiritually equip them for their future service. Others are seeking a relationship with the living God and find him through OCF, or another Christian ministry at the Academy, or the Gospel preaching of one of our great chaplains. Then through Bible study, discipleship, retreats, conferences, and mission/service trips, we provide multiple venues to mature in and share their Christian faith. Before wrapping up this amazing faith story, I want to share the brief yet powerful testimonies of several of the cadets in our ministry.

Here is one cadet's e-mail to me shortly after he became a believer in Jesus Christ at our winter retreat:

> *Basically I've always had a fear of flying and my fear was of dying and also not even knowing where I was going when I died. It was to the point where I was even superstitious; I had to have things a certain way before I flew. During the Winter OCF retreat, on February 19, 2005, I took the next step in my faith and fully gave my life over to God.*
>
> *This spring break was my first time flying since then and things were not going like they usually do. Surprisingly I was not worried. I had prayed the*

*night before about our safety on our journey. The day of the flight, even though nothing was going right, I had a calm feeling come over me, and I was not scared to get on the plane. I felt that even if the plane went down I knew where I was going.*

*So I made the most out of the plane flight and looked out of the window as we flew — for the first time in a long time — and enjoyed the awesome creations from God. It was the best experience flying that I ever had. I felt that God definitely touched me as I flew and let me know that I did not need to be scared and that I had nothing to worry about.*

This cadet completed a 15-week discipleship program (outlined in Appendix 1) with me to grow deeper in his faith and to be able to teach others the same truths. He was a "God's Gang" leader/mentor at West Point with the community youth. His friend teamed up with him as a small group leader and she, also, completed the discipleship program and became a "God's Gang" mentor. They have faithfully carried on Jon Shine's legacy by investing their lives in younger believers who want to become mature in their faith.

Another cadet was the discipleship team leader in our OCF ministry during his Firstie year and has gone through the discipleship program. He helped others who needed assistance to come to faith in Christ by encouraging them, teaching them, and leading them to a deeper faith walk.

Here is an e-mail he sent me during that time:

> *Since sophomore year, one of my best friends has been going down the wrong track. He has favored worldly things and drinking over other aspects of his life. He is unhappy and you can tell. As a senior, I am his roommate again and tried to answer his questions about God. He was to the point of wondering whether God even existed anymore. I gave him a copy of Mark Cahill's book,* **One Heartbeat Away.** *I told him that it would help him get a better perspective on God. For the next two days, I would see him reading the book every time I came into the room...God works in miraculous ways.*

I asked this cadet about his prayer life following a teaching session we had on the subject. Here are his thoughts, from which all believers can learn:

- <u>You can pray at any time.</u>

> *If you feel flustered during the day, just pray! I always thought that you should only pray in church or in a Christian setting. Yet, praying to God during the day will give you a closer relationship to Him. Pray while you are going to class. Ask God for help before lunch. Finding comfort in God throughout the day has helped me keep a positive spirit and grow closer to Him.*

- <u>Pray before you start your day every morning</u>.

*Mark 1:35 (NIV): "Very early in the morning, while it was still dark, Jesus got up, left the house and went off to a solitary place, where he prayed."*

*I had never taken the time to get up early every morning to pray and spend time with the Lord. I have never been happier, as I have started to give thirty minutes of my time to God every morning. It doesn't matter how long you give Him, as long as you sacrifice some time for Him in the morning.*

- <u>Spend some hard time in prayer</u>.

*Over the weekend, when you have time, go to a place of special meaning and pray. Give some real time to the Lord — time when you don't have many requirements — and spend some quality time with Him. Follow His example, as He prayed all night (Luke 6:12). Although you might not pray all night, the weekend is the time you can really give your time to Him. The weekend is our free time. How much of it are we giving to Him, if we are even giving any at all?*

This cadet and another in his company started a spiritual mentorship program within their cadet company, recruiting many of the underclassmen and women who desired to be mentored, and building a core team of older cadets who desired to be mentors. This particular company-level (about 100 cadets) ministry was an outgrowth of how Jon Shine mentored me and transformed my life in the process.

The cadet's letter sent to prospective participants in this ministry said the following:

> *Another reason why this is so important—the leader of Officers' Christian Fellowship here has been working with me and when he was a plebe, back in the day, his company commander (Jon Shine) asked him to join a mentorship group and that was the reason he got so excited about and turned his life over to Christ. And now he impacts numerous lives here at West Point.*

The Firehouse Faithful (the company's mentorship program) are at it even now, growing in their faith and helping others become disciples, too. These young men and women represent only a fraction of the faithful cadets on fire for Christ at the U.S. Military Academy today. They want to serve their country during one of the most dangerous and uncertain times in our nation's history. They also want to serve Jesus Christ as ambassadors for Him in uniform.

To illustrate the power of discipleship on succeeding generations of believers at West Point, here is a testimonial from a young cadet who was the spiritual mentee of an OCF ministry senior (from the Class of 2007):

*We get together usually once a week to discuss the discipleship plan outline (the one with all the weeks laid out for progress checks). Whenever we meet, it's either on the weekend or at dinner, and we always discuss how my progress has been holding up and things I can improve upon.*

*He has also shared with me techniques on keeping myself personally accountable for things I would like to improve upon in my life spiritually.*

*He has also shared websites with me that I can go to in order to seek help in my personal studies.*

*We discuss particular chapters such as in Matthew, and I tell him what the passage means to me. Everything I have sought in discipleship, I have received.*

One final note about Jon Shine's spiritual legacy: While at Burke Community Church in Northern Virginia during the time I was a civilian corporate consultant, before going to West Point with Barb as the OCF staff representative, I was inspired by Jon and supported by Pastors (Dr.) Jack Elwood and Mark Green to start, in the fall of 2004, a discipleship program that took four faithful men (mostly my peers) through a challenging 33-week spiritual development plan. As of the fall of 2006, there are three generations of men and women who are products of that program, helping the new discipleship pastor bring more into training.

Jon, I'm sure, would not want us to focus on him or write a book about him. He would much rather we focus on the

Lord Jesus Christ, giving Him the credit for spiritual success when any credit is deserved. True believers, however, know that faithful disciples of Christ who have gone on from this life, have always left behind them powerfully vivid signs—trail markers—that point the way toward Him. One of those trail markers Jon left us was an example of total reliance on the Holy Spirit, our Comforter and true Friend. Robert "Doc" Jackson, the brave Army medic who heard Jon Shine's last words, now understands why Jon's words— coming from his platoon leader lying five feet from him with a serious wound to his head and enemy fire all around—were so amazingly calm and full of assurance and peace. Jon embraced the Holy Spirit all of his life and that Spirit's peace—a peace that "passes all understanding"- comforted and embraced Jon in his last moments on earth.

In F.B Meyer's book, *David*, the author traces this young warrior David's life through the steps by which he became king...those steps in which his character was formed. In one of the book's poignant passages, he describes David, who, without hesitation, bolts toward the enemy, Goliath, with great valor and skill when that enemy threatened his men. Meyer describes this soon-to-be warrior-king, and by inference all those who emulate him, as those "in whose breasts the dove-like Spirit has found an abiding place, and whose hearts are 'sentineled' by the peace of God...these are they who bear themselves as heroes in the fight."[2] I thought about Jon and his heroic sacrifice the moment I read this passage.

[2] F.B. Meyer, *David*, London, 1953, Preface and p. 34

The path we tread as we move toward that prize of God's high calling—our reigning with Him in glory throughout all eternity—is well lit by the courage and character of a young Army lieutenant whose heart was focused on Christ-like servant-leadership and whose calm, peaceful voice and bright, engaging smile remain fixed in our memories and encourage us on… every step of the way. And as our awesome heavenly Father lifted Jon out of his final valley of death and into His holy presence, He will take us through all our valleys and lift us out by His mighty power and love. His grace is certainly sufficient for each of us. His strength is made perfect in our weakness.

Thank you, Jon, for the eternal legacy you have left us. We pray for the strength, wisdom, and inspiration to carry it on.

# Chapter Five Roll-up

## Principles from Jon's Spiritual Legacy

- Generations of disciples of Jesus Christ can be born out of the faithfulness of one committed man or woman.

- Spiritual mentorship and discipleship are the result of committing to the Lord's command to us all—"go and make disciples"—which isn't just the work of overseas missionaries.

- Bible study is a powerful key to effective Christian living, but not the sole key. Training others *all* the things Christ taught His disciples and all of us, today, through His Word, is the heart and soul of effectively talking the Christian talk and walking the Christian walk.

## "Harder Right" Choices

- Jon Shine's life and death inspired the author to work with two younger cadets at West Point, both of whom were willing to sacrifice their free time to go deeper spiritually and to risk negative and derogatory comments and reports or rumors by peers and seniors.

- The author shared his personal faith story with his rifle platoon and mentored a young enlisted soldier who came to him and desired to grow in his new Christian faith, not an easy or popular thing to do in the '70s...or now.

- The author's wife and children became the highest priority in his life, after his relationship with Christ, and the author's relationship with his mentor, Jon Shine, impacted them powerfully.

## Putting and Keeping Christ at the Center

- The author felt compelled—by the Holy Spirit, by the example of his spiritual mentor Jon Shine, and by the world circumstances that almost caused him to deploy into combat—to share his faith with his rifle platoon. The result was a powerful discipling relationship with a young soldier who had accepted Christ as his Lord. This kind of sharing is best done after duty hours, not in uniform, with no coercion using rank or position, and in accordance with local military protocol.

- Working with a youth group while assigned in Panama, the author and his wife were able to naturally develop close personal relationships with a number of the youth who desired to grow deeper in their faith and allowed them to stay focused on Christ as the center of their lives, and help others experience that same victory.

- A young man, whom the author mentored, became a Christ-centered chaplain in the Army and had a powerful impact on dozens of young soldiers, including a sergeant who is carrying on the Lord's work within the Army, focused on sharing the love and gospel of Jesus.

Discussion Questions

- What valleys did the author experience while on active duty in the Army, and how did he climb out of them?

- How did the author integrate his faith into the military profession, as a cadet and as an officer?

- How will you try to integrate your Christian faith into your school experience, job, or profession?

- Is it feasible to engage in a discipleship or spiritual mentorship program while fully employed in a job or profession that is time-consuming and/or with a family that requires much of your time? Why or why not? Why and how did the author try to do this? What motivated and inspired him?

- How do you plan to develop relationships with others with whom you work and who may desire to grow in their faith and allow you to mentor them? Specifically, whom do you know, by name, who you might spiritually mentor?

## Jon's Spiritual Legacy

- *West Point Cadets:* Jim Blackwell, Greg Schumacher, and Art Hill
- *Other military servicemen:* Specialist Fred, Timothy Mallard, John K., Geoff D. (later a member of Congress), Suat G., Oscar A., Mark I., and Cecil S.
- *Cadet Classmates of Barry Willey:* Ron Hawthorne, Jim Hougnon
- *Their family members:* Jonathan Hougnon, Jonathan Willey, Rachael Willey, Barbara Willey
- *Other West Point Cadets who were discipled:* Brian B., Tracy M., Michael N., Jim F., Curt B., Ryan F., Peter C., Eddie M., Drew M., Edward M., Sean F., Andy B., Eric W., Nick P., Joel H., Ian M., Paul G., Ray B., Matt P., Caleb G., Mary Alice P., Lucas M., Chris G., Daniel N., Jesus G., Sean I., and Nathan M.

- *Mentored at Burke Community Church:* Pete W., John P., Dave W., Tom B., Paul G., Tim L., Carlos K. and Mike and Stephanie M.

- *Cadets in Timothy Club when Barb and Barry were on staff at West Point:* Mike P., Glen K., Scott M., Jim B., Nate P., Les K., and Kevin B.

- *Discipled at Grace Church in Florida:* Paul, Jimmy, Tim, and Chip

- *Discipled as high schoolers in Panama:* Kent S. and Matt U.

These have all sought to be faithful followers of Christ throughout their lives, many taking steps to go deeper, invest in others, and build a spiritual legacy for the kingdom of God, as Jon Shine so faithfully did. They are his eternal legacy.

> "You yourselves are our letter, written on our hearts, known and read by everybody. You show that you are a letter from Christ, the result of our ministry, written not with ink but with the Spirit of the living God, not on tablets of stone but on tablets of human hearts." 2 Corinthians 3:2-3 (NIV)

*****

*Out of the Valley*

*"We loved you so much that we were delighted to share with you not only the gospel of God but our lives as well, because you had become so dear to us."*
*1 Thessalonians 2:8 (NIV)*

# Appendix 1
# Discipleship Plan Outline

The author is using this program to spiritually disciple/mentor young men and women at West Point in order to help them deepen their Christian faith-walk and grow stronger in their relationship with, and love for, Christ. It is carrying on the powerful, eternal legacy that Jon Shine left at West Point. It is only an outline and can be adapted and adjusted by anyone who may want to use it as a guide to teaching believers, young or old, the basics of the Christian disciplines. It is fifteen weeks because that works well for cadets at West Point, or perhaps students in a university setting—about one term long. Each section can be expanded, as needed, to go deeper into the disciplines of becoming a disciple of Jesus Christ. Ideally a teacher or small group leader could read through this five-chapter book with his disciple or small group, taking one chapter a week, and pausing to discuss the questions at the end of each chapter. After completing the book, the leader could then begin the 15-week program.

It will change lives!

## First Three Weeks (meet once per week)

<u>Following Jesus… what does that mean?</u>
- Research the Gospels and record at least five references where Jesus talked about what it meant to be His follower. Focus on the Sermon on the Mount (Matthew 5–7).

- Meet and discuss these traits of a disciple.

*Action Steps:*
- *First week*: In what specific area are you strongest… why?

- *Second week:* In what area do you need the most work… why?

- *Third week*: What specific behavioral change will you commit to make… and when… to be a more effective disciple of Christ?

## Second Three Weeks
The Word of God

<u>First week:</u>
- Study how Jesus knew the Word and used it powerfully in His ministry—during temptation (Matthew 4); as a young boy in the temple (Luke 2:39-52…implication here is that Jesus spent lots of time

listening to the Word read and taught as a boy so that when He was in the temple amongst His elders at age twelve, they were astonished at His understanding); when He read from Isaiah and declared the prophesy fulfilled in the reading (Luke 4:16–21...here we see Jesus knowing well the Old Testament book of Isaiah, turning immediately to the text that He read that was the fulfillment of the prophesy). Bottom line: Jesus had spent many days of his youth immersed in the Scriptures...reading it, hearing it, memorizing it, and understanding it. From this point, focus on having a regular Quiet Time...use any program that will work for the disciple...a good devotional like Officers' Christian Fellowship's *100 Days Bible Study* (as a devotional tool) or *My Utmost for His Highest* or the Psalms or Proverbs.

- *Action Step:* Commit to a one-month Quiet Time...if they stay with it, they will want to continue beyond a month.

Second week:
- Focus on Bible Study and Bible reading. Use the Navigator's *NEW 2:7 Series* or OCF's *100 Days Bible Study* as a tool (found online). Or go to page 42 of this book and use Jon Shine's favorite study technique—the inductive method. Then get a read-through-the-Bible-in-a-year pamphlet (easily found online) and discuss the importance of such reading.

- *Action Step*: Commit to doing a Bible study a week for a month OR starting a read-through-the-Bible-in-a-year program. (They can try both if inspired and motivated but be careful they follow through and stay with it.)

Third week:
- Bible memorization. Psalm 119:9–11 is a foundational passage on this topic. Look at it and talk about it. Discuss memorization and its importance in the life of a disciple of Christ. Share examples of how the Holy Spirit has brought verses to mind when you needed them. Share some of the key verses you have memorized. Why have they been important to you?

- *Action Step:* Memorize the following verses on assurances… John 16:24 (prayer); 1 John 5:11–13 (salvation); 1 John 1:9 (forgiveness); 1 Corinthians 10:13 (victory over temptation); Proverbs 3:5–6 (guidance). Memorize one every two weeks during this program.

## Third Three Weeks
Prayer

First week:
- How did Jesus demonstrate the act of prayer? You can search the internet using "Jesus Prayed" or go to www.believers.org (click on "Jesus Prayed" link) and

find a great compilation of how and when He prayed from the Gospels. Study and discuss this... it's our best model of how we should do it.

- *Action Step:* Develop an evangelism prayer list on a 3" x 5" card. List the names of a few friends or people with whom you've developed a relationship, and start praying for opportunities to share your faith story and the gospel with them. Start praying!

<u>Second week</u>:
- Study John 17 and discover Jesus' passion for intercessory prayer. He prayed for His disciples... but also for those His disciples would reach with the gospel.

- *Action Step:* Develop an intercessory prayer list on a 3" x 5" card and include all those you should be praying for regularly...family, friends in Christ, pastor, mentors, those you mentor, those they will reach for Christ... add to this list, as needed. Start praying!

<u>Third week</u>:
- Prepare a time in your schedules to spend a half day in prayer. The best time is usually Saturday morning, starting early and going about five hours. Gather together first, then break out and go to a quiet, solitary place. Spend time in the Word and pray

through a Psalm. Have a list and use it to jog your thoughts. Sing praises; just enjoy sweet fellowship with the Lord. Then reconvene and share. It usually is a life-changer.

## Fourth Three Weeks
Sharing Our Faith in Christ

<u>First Week</u>:
- Develop your faith story... your Christian testimony... have disciple write it down in a "3-C" format:

    o Concise (one page typed or a few 3" x 5" note cards)

    o Compelling (interesting... what happened to you that would make someone want to listen to you)

    o Christ-centered (yes...it's YOUR personal story, but the real focus is Him... include one Scripture that describes what happened to you, e.g., John 3:16)

- *Action Step:* Practice this on each other or with Christian friends several times in the next week.

Second Week:
- Learn a clear and effective technique of presenting the gospel to a non-believer...e.g., *The Bridge, The Roman Road, The Four Spiritual Laws*.

- *Action Step:* Practice this technique on a Christian friend in the next week.

Third Week:
- Work on presenting both testimony and gospel at the same time, in brief time period (four minutes or less). If you have thirty minutes, take it! But in our fast-paced lives, it's necessary to be able to present your story and the gospel briefly, yet clearly and powerfully. Take turns, trade off, and work this hard.

- *Action Step:* Practice this with Christian friends this week.

## Fifth Three Weeks
Evangelism Practical Exercise... Tying it All Together

Each week, for three weeks:
- Meet to pray about engaging someone on your evangelism prayer list. Discuss opportunities that might present themselves that week to share your faith story and gospel presentation.

Expect God to provide that opportunity... **He will!**

- *Action Step:* Engage someone on your evangelism prayer list following your session each week and share your faith story and the gospel with him or her. As God leads them to a decision for Christ, be prepared to take them through this discipleship program.

# Appendix 2
# Testimonials

Many men and women have been touched either directly or indirectly by Jon Shine's life and commitment to Christ. Today their impact on others continues Jon's legacy as their lives bear positive personal, professional, and spiritual fruit.

## Testimony of Rachael Elizabeth Willey

I remember Jon Shine's name being said a lot while I was growing up. I am sure it was talked about before I was even aware, due to the fact that my brother is named after him. Although I never had the privilege to know Jon, I do know him through my dad.

I made the decision to give my life to Christ when I was eight years old. It has been the greatest decision I have ever made. Before that time, I was aware of who Jesus Christ was and knew my parents' testimonies, but it was not until I made the decision myself to follow Him that my life changed.

I know that God had this planned out even before Jon, my dad, or I was born. Who would have known the impact that one comment would have made and how many lives that one comment would affect? Jon asking my dad to a Bible study at West Point was the beginning of the domino effect of how one person's faith affects another and that person's affects another and so on.

Jon mentored my dad, helped him come to "his own" faith and invested ten months of his life to help my dad and other men grow deeper in their faith. Because of that leadership, my dad took what he had learned from Jon and applied it to his life and taught me what it means to be a Christian, encouraged me to develop "my own" faith, invested 26 years of his life to help me grow deeper in my faith, and has been one of two mentors in my life—the other being my mom. As I have devoted my life to the work of Christ, I have asked God where I can be a mentor and God has given my heart the desire for the poor, oppressed, and helpless.

God's plan is perfect and now I am mentoring children in my work as a nanny and as a Sunday school teacher. My degree in social work has given me the opportunity to work with the poor, helpless, and oppressed. I am now an elementary school teacher in an inner city school where I am a mentor and role model for my students.

As I pray for my future husband and children, if that is the Lord's will, I pray that I will encourage them in their faith, be a mentor to my kids, and be the kind of godly wife and mother that is described in Proverbs 31. And, Lord-willing, the domino effect will continue. I thank the Lord for Jon Shine and placing him in my dad's life at the perfect time and in the lives of so many others.

*Out of the Valley*

## Jonathan David Willey's Testimony

Throughout my life I heard about Jon Shine and the lasting impressions he made on my father, our family, and on many other families who knew Jon personally. Not until recently has Jon's life made such an impact on mine. I grew up in a strong Christian family where my parents instilled in me the knowledge of the Bible and the power of Christ's love for me and our whole family. Along with this love for the Lord I have also learned to respect my country and be proud of those who serve diligently to protect and defend this great land.

I have seen many good men and women wear the military uniform and I am proud to have a father who does it with vigor and zest that is not far surpassed by anyone in this day and age. I never got the chance to know Jonathan Cameron Shine but I feel that he is embodied today in the actions and faithfulness of my father. Jon made such an impact on my dad that I was named after him.

I admire Jon immensely for his courage and bravery in the face of danger. I know, from hearing my father preach and talk candidly to me about Jon, that Jon was a good man, a great husband, and a devoted soldier. My father is all of these.

To this day I keep a flag hanging behind my bed to remind me of those men and women, like Jon and my father, who risk their lives to make this country safe for me. I can't repay them for their bravery; I can only hope to honor them through the way I represent my country to others. I have a respect for my father and for Jon Shine that far surpasses any feeling I have for anyone.

The man I am today was created by God, but has been continually influenced by the model that my parents have set for me. Even though Jon Shine never met me while he was alive, he knows who I am. Someday I will know him and will be able to thank him myself. But until then, I will continue to follow the path laid out for me by the walk of my father, a true, good, and faithful servant.

## Barbara J. Willey's Letter to Jon Shine

Dear Jonathan,

We never had the opportunity to meet, but for a few moments I want to share with you and thank you for so many things. I do wish I could have met you. When I met Barry in the summer of 1969 you had already graduated. You know how guys are, especially when they are 18 and rather shy, they don't really talk much about their friends, so I didn't really get to know much about you that summer, but I now realize that you were there. You were there in the influence you had on Barry. Barry was very different from the other guys I had dated. He was not just a "Christian guy." I had really only dated Christian guys since high school, but Barry was different. He was serious about his faith...so credible. He didn't just "talk" about the Lord, or participate in church, or attend Bible study, he really lived it. It impressed me.

I didn't really trust men because I had grown up in a home with a father who was an alcoholic (praise God he became sober and was for 30 years...and is now there with you and His heavenly Father), so I was afraid to trust. I just had never met anyone quite like Barry. He was so polite and gentle. His kindness was evident to everyone who knew him, and he was kind to me all of the time. He treated me with the greatest of respect and was always interested in what I was doing. This may not seem unusual to you, but for me this was HUGE!

You see, Jon, Barry and I are as different as they come. He is predictable and calm and I am impulsive and, well let's just say, not exactly calm. I am about as extroverted as they come, and you know Barry is rather introverted. We are very different in our personalities, but in our values, in our love for the Lord, in the things that matter when it comes to building a marriage, in those things we are most compatible.

When I would go to West Point to see Barry, I often stayed with Al and Sandra. Maybe God was allowing me to know you a little by knowing your brother. Al and Sandra were great. I was able to see a Christian marriage and family in action. When Barry and I married in June of 1974, the girls from our wedding party stayed with Al and Sandra. It was wonderful to have them in our lives. I know now that for Barry it was like having you there, in a way.

We were back at West Point ten months after we married, as Barry was selected to be the Aide to the Commandant. Al and Sandra were still there and we learned much from them. I was pregnant with our first child and our daughter Rachael was born on Veteran's Day in 1975! We were so proud. It wasn't long after that that we decided to have another child and I became pregnant again. During this time I was the Spiritual Life Chairman for our Chapel Women's group. I had heard from someone that your mother was traveling and speaking to women's groups about dealing with grief. By then your dear brother Tony, a skilled, courageous, and dedicated Air Force fighter pilot, had been missing in action. It seemed so right to have her come and speak while we were at the Academy. She agreed.

I will always remember what a joy it was to hear her. Joy doesn't sound like the right word to some, perhaps, but to me it was. She shared the joy of knowing that your loved ones are with the Lord. You would have been so proud of her, Jon. Everyone there was able to learn a little bit about you and your brother, and we all were so blessed to hear your mother share. We knew that any of us in that room could be the next mother or wife to grieve over the loss of our "special soldier" so we listened well!

After the program we had your mother to our home for lunch. I was quite pregnant by then and we shared with her that if our child was a boy, we had decided long ago to name our son Jonathan, after you! Your mom was so thrilled that we had made that decision. Soon after she left and we were on our way to Fort Benning, Georgia, on July 8, 1977, our son Jonathan David Willey was born. It was a very exciting day for us. We had so wanted a son and he came screaming in at ten pounds of energy.

Later that month I sent a note to your mother telling her of the news. It was to our surprise that she wrote back and shared with us that July 8 was her birthday, and you were expected to be born on that day, but came a day early! Isn't God good? He gave our family a Jonathan on the day you were scheduled to be born, and on your mother's birthday.

Your influence lives on now in our son. He is a wonderful young man. We took him to D.C. when he was in elementary school and for the first time we told him about his name's sake. We stood at the Wall and Barry shared with him how proud he should always be for being named after such a man as you. How you had always put God first in your life. How you were a man of character and integrity. We shared with our young son that it is important to know about your name, and that a name is something God honors. Jon, you can be proud of this young man. He is married to our dear Jamie; they both love Jesus, and plan to have children. He bears your name well, and both he and our daughter, Rachael, share with others the powerful impact you have had on their dad and on them. I guess what I really want to do, Jon, is say thank you. Thank you for listening to that dear custodian in the field house those many years ago when he shared the gospel with you. Thank you for taking the time for Barry, a mere plebe, and for caring enough to disciple him. Thank you for serving our country in a war that no one really wanted and for which no one said "Thank you" at the time. And most of all, Jon, thank you for being willing to give your all, and in doing that, leave a legacy that lives on even now in the lives of those you have touched.

I wish we had met, but wishing isn't the important thing. The good news is we will meet. I love the verses in John when Jesus tells His disciples that He is going to prepare a place for them, and they are not to be sad, because He is going to come back and get them, and take them to be with Him and they will be together forever and ever…

Jon, maybe we can all be neighbors some day!

With anticipation and gratitude I remain,
Barbara

## Testimony of the Casualty Notification Officer for the Shine Family—William "Bill" Tozer

In the fall of 1967, I had already completed one tour in Vietnam and would soon graduate from the University of Southern California with the Master's Degree in Mechanical Engineering that I needed to prepare me for a tour on the faculty at West Point. Those of us on active duty at that time can probably empathize with the mixed emotions pulling at my heart. Graduate school in Southern California was good duty. Plenty of family time, double-majoring in handball, predictable days, civvies, no IG inspections, weeks off between semesters, and, ….., and …., and a persistent sense and burden that I wasn't carrying my share of the load in Vietnam. So, armed with the fact that I would have my degree in hand half a year before I was due at West Point, I confidently called my assignment officer at the Pentagon and asked that I be sent back to Vietnam for those 6 months.

## Out of the Valley

He politely told me that I was untouchable from his point of view. West Point was paying for my education and for that reason I belonged to them. They would not agree to such a plan. Well, I tried.

Then came Tet in the Spring of '68, and with it an even greater concern, perhaps even fear is a better word, that I just was not doing my share. Of course the feelings were all mixed up. Nobody wants to leave his family for a year. Nobody in his right mind wants to go to war, especially after he has actually experienced it. But again, there was nothing I could do about it for a while. On to West Point I went and promptly got immersed in a new and rewarding job with its own challenges.

Life there was even better than in Los Angeles. I didn't have to live among all those silly civilians. I much preferred the company of the military community; if we absent-mindedly left the keys in the car, no biggie; if when lying in bed at night I remembered that I had not locked the door, no biggie; the handball courts were even closer than in Los Angeles and I could even go skiing a couple of miles from my house. The hunting and fishing were great and the social scene unhealthily robust. But I was also getting to see on a first-hand basis the havoc being wreaked on the hopes and dreams of those families whose husbands and fathers, just a few months ago my peers on the faculty, were now coming back to their alma mater to be buried alongside the heroes of previous wars. All those tears and all that anguish and all that heartache just intensified the war in my heart over "not doing my share." Then came a notice in the post daily bulletin or some such medium that they were looking for volunteers to be casualty notification and assistance officers. I jumped at the chance to carry a little more of the load.

My turn came on a Saturday morning in October of 1970. I received a call from the West Point Casualty Office that I was to go to Pleasantville, New York, to notify the wife of First Lieutenant Jonathan Shine that he had been killed in action in Vietnam. I was given a detailed list on the proper protocol and was told that an Army sedan would be by to pick me up shortly. At the top of the list were instructions to take a chaplain from West Point along with me, or if that were not possible, conscript one from the local community, but by all means don't go there by myself. I don't know how many chaplains were assigned to West Point at that time, but not a single one of the half-dozen or so names on the list were available. I knew nothing about the circumstances of death or the man. I did not even know that he was a West Point graduate. Although I considered myself a religious man at the time, I was not a Christian and waded into this assignment in my own strength and wisdom, both of which were soon to be discovered woefully lacking.

My driver and I were in Pleasantville within an hour of leaving West Point and I began asking people I could find on the street directions to the address I was given. None recognized it. I went to the police station and found that even they were uncertain, but the desk sergeant suggested that I look in a relatively new housing development at the end of such-and-such street, to which street they gave me directions.

*Out of the Valley*

Soon my driver and I were passing a church, whose denomination I cannot remember and at the time it did not matter. I told my driver to pull in hoping I might find a pastor to accompany me on my call. It was a fairly large structure with a few cars in its large parking area so my hopes on getting help were raised. I found an open door, entered into a hallway lined with offices and could hear a choir practicing someplace else in the building. I don't recall what they were singing but it was beautiful; it echoed in the hallway and gave me a sense of peace that things were going to turn out well. I started walking toward the sound and quickly encountered a woman that I was certain could have been a twin sister of Sue Pellicci, wife of my classmate and handball partner Jack Pellicci, also currently at West Point. I was as taken aback as she. She asked how she could help. I told her I needed to talk to the pastor. She replied that he was out of town until later on that evening. I asked if she knew how to get to the address I showed her. I needed to talk with a Gail Shine at that address.

Things began to unravel quickly. "Oh yes," she cheerily replied. "Gail is staying there with her husband's folks while he is in Vietnam. She worships at this church. But Gail is not home right now. She is at Fort Rucker, Alabama, attending the funeral of her husband's close friend and roommate as a cadet at West Point, who was recently killed in Vietnam." I don't know what my face looked like but I know it wasn't encouraging. Just then she looked all up and down at my uniform and began to connect the dots. With a suddenly worried and frightened look on her face she asked, "Is everything OK?" I have no idea how I responded but I know I did not tell her the truth. I suspect my story wasn't very convincing, either. How could I have possibly prepared myself for that encounter? I never learned her name, but she was the first in Pleasantville to know.

I was in a stupor by the time I got to the Shine residence. Right at the top of the list of instructions was "Do not under any circumstances tell anyone but the widow!" and here I was ringing the doorbell of Jon's mother and father without a chaplain and knowing the widow was over a thousand miles away. I may even have said a prayer.

A stately woman, clearly older than I, answered the door. She was visibly shaken before either of us said a word. I asked if she was Jon Shine's mother. "Yes" was the nervous reply. I said, "I need to talk with you. I have some very bad news." If my memory serves me, she got wobbly and I escorted her to the nearest chair I could find. I think we were in the kitchen. I don't know what I said or how I said it. I did my best to be gentle and compassionate. And I somehow got the message across. Jon has been killed in action.

She quickly got up from the chair and ran down the hallway calling out to her husband "Jon has been killed! Jon has been killed!" I saw him briefly as he put his arm around his sobbing wife and they disappeared behind a closed door for what seemed like an eternity.

When they reappeared, both red-eyed from crying and cheeks still wet, they had a composure about them that surprised and puzzled me. I didn't dwell on it much at the time but hindsight provides the explanation. Their first words were to the effect that Jon was in a safe place. It's OK with him. He's with the Lord. But we need to get the news to Gail and take care of her. Mrs. Shine looked at her watch and said something to the effect that they couldn't reach Gail right now, she was at a funeral. I think I told them that I knew that and I needed to be the one to tell her. So I would wait around until they could get her on the phone so that I could tell her. They pleaded and pleaded with me to let them be the ones to tell her. I tried to explain that I had some orders to comply with but their obvious love and compassion for Gail persuaded me once again into noncompliance. I told them I had no details concerning the circumstance of Jon's death but as soon as I received any I would be back with the details, including when Jon's remains would arrive at Dover Air Force Base in Delaware.

Then some more inexplicable coincidences began to emerge. Jon had an older brother, Al, who was also a West Point graduate, who was attending Harvard University in preparation for his teaching tour, but who was in town this particular weekend on the occasion of his wife's college reunion. They were presently at one of the reunion activities and we needed to get the word to him. Further, Jon had a sister, Sally, who was also in the Army currently attending Columbia University's nursing program not very far away in New York City. They had to get the word to her as well. Stupor was being replaced by awe at the highly unlikely circumstances cascading before me.

Then, Jon's mom and dad turned to me and said something like, "Major Tozer, you must have the nastiest job in the Army. Is there anything we can do for you? Can we get you some cookies and milk?" Before I could pull together an answer, cookies and milk were on the coffee table in front of me.

I was probably there at the house for less than ninety minutes and when I left my head was spinning. I couldn't even begin to digest what had happened. Jon's parents had agreed to call me at West Point as soon as they had notified Gail. When I returned to West Point to report to the Casualty Office all that didn't happen the way it was supposed to, I was actually fearful of receiving disciplinary action. The Casualty Officer was greatly displeased with my report and directed me to be back at Pleasantville the moment Gail returned to personally notify her. I called Jon's mom and dad, told them of my orders and asked them to let me know when she would be arriving. Again they pleaded with me, successfully, to let them handle it. They promised to have Gail call me when she arrived so that I could speak to her myself. Many hours later a very tired Gail called me in the middle of the night to let me know that she was back in Pleasantville and knew of Jon's demise. I called the Casualty Office and gave some kind of mealy-mouthed report that I had personally spoken with Gail concerning her husband's death. Fortunately, they didn't ask any more difficult questions.

But there remained a whole lot of unanswered questions concerning the circumstances of Jon's death. Over the next many hours I was able to repay the Casualty Office in kind by hounding them for more information to pass on to the

Shines. How, where, and exactly when did it happen? Where are the remains? When are they due at Dover? What are my next duties as the Survivor's Assistance Officer?

Perhaps twelve hours later I got a little piece of information. I forget now what it was but I hurried back down to Pleasantville to pass it on and to begin the duties of Survivor Assistance Officer. Surprise! Jon had another brother, Tony, who was a fighter pilot also serving in Southeast Asia. He had already joined up with Jon's casket in Saigon and was now escorting it back to Dover. He would arrive at such and such a time and arrangements had already been made for a hearse to carry Jon's casket to a funeral home in Highland Falls, just outside the gate at West Point. Al was going to be the Survivor's Assistance Officer and was overseeing Jon's return and burial at West Point. They knew much more than I did and had everything under control. They would let me know when the casket had arrived in Highland Falls. I went from being the coordinator to a spectator. I was much relieved but felt like I had not been much help.

About a day later I got a call that the casket was at the funeral home. When I got there I encountered another unforgettable sight. The flag-covered casket was in a poorly lit room illuminated by only one window. Beside that casket, standing at parade rest, was an Air Force Captain who looked like he had not slept for a week. Every button on his uniform, his rank insignia, and his shoes glistened in the dim light. He kept his eyes straight ahead and paid me no attention. I am a relatively big man but it seemed to me that he was bigger than life and I knew right then I had better not touch that casket.

Funeral services were soon held at the Old Post Chapel close to the cemetery and for reasons I cannot recall all these years later, I did not attend. I heard later from someone who did attend that the chapel had been filled to overflowing with General Officers and Colonels and officers of all rank and sort, and by post personnel, and cadets all paying homage to this man who less than 18 months before had been a cadet himself. He had graduated while I was there and I had never heard of him. What made him so special to all these people? I still know only a small part of it. But I heard that as a cadet, he had been in charge of all the Sunday school classes for all of the children on post. He had held a high position in the cadet chain of command.

Time went by quickly after that. I knew I would be returning to Vietnam in a few months and for reasons explainable by only God, my wife and I began to get involved in some officers' Bible studies. Howie and Gracie Graves [a fellow Academy teacher and later Superintendent of West Point] were among them, and there were many others whose names I have forgotten. In hindsight I can now recognize they were siccing the hound dogs of heaven after me and my wife as we both soon traded our sins for the grace of God at the foot of His cross.

*Out of the Valley*

*The four Shine kids, left to right, Jon, Sallie, Al, and Tony
All served in the Vietnam War, Al being wounded, and Jon and
Tony being killed in action. Sallie served in the Army Nurse Corps
and was a Red Cross volunteer in Vietnam.
This photo was taken at a family reunion at Black Lake, New York,
in May of 1970, the last time they were all together.*

I survived another year in Vietnam and returned to Fort Carson, Colorado, for a couple of years. While at Fort Carson, my family and I even spent a weekend at the Officers' Christian Fellowship retreat center in Spring Canyon, Colorado. When we signed in we were assigned to the Jon Shine Lodge! After Fort Carson we got posted to Command and General Staff College at Fort Leavenworth, Kansas, in the summer of 1974. The years between West Point and Fort Leavenworth were kind of a blur and somehow the memories of sadness and grief either got crowded out by my own personal concerns or as some kind of a subconscious self-defense mechanism.

Then, as only God could arrange it, my wife and I got involved in some more Bible studies at Fort Leavenworth. They were good for my soul for I was not the man God wanted me to be and He was making me very much aware of it. One evening we walked into a Bible study gathering and my eyes met those of a man I knew I should remember but I couldn't place him. From the way he looked, I felt he was thinking the same of me. During the fellowship time after the study was over, we got reacquainted. It was Al Shine! Wow, what a small world. We weren't into our conversation very long when I asked about Tony. Memories of my very brief encounter with him were even more fixed in my mind than my recollection of Al. "Where's Tony?" I asked. "Boy, I'll never forget that man. How's he doing?"

"Bill, Tony got shot down over Laos last year and he is missing in action."

It floored me. I don't know whether my heart went up to my throat or down to my toes. I could hardly believe my ears. The look on Al's face assured me he was not joking.

Years passed. Good and not so good things happened. I had become a born again Christian right after my tour at West Point was completed. However, I remained a spiritual baby for a decade after that. No growth. Too many parties. Too little Jesus. Divorce. Turmoil. Estrangement from my children. I think I actually saw it coming while we were still on the faculty at West Point. I accept full responsibility.

Remarriage. Retirement from the Army and a move to Moscow, Idaho, to get an MBA and start life over again. My West Point roommate, Harold Smith, was here. God was not giving up on me.

He led me, my new wife, and her two children to a Nazarene Church here in Moscow. It felt like it was the first time I had ever seen people actually worshipping God instead of singing hymns. It was glorious! We saw people who were probably very much like those I had encountered at West Point and Fort Leavenworth Bible studies, but my eyes were dim back then. We learned about the lordship of Jesus Christ. We were growing spiritually by leaps and bounds.

But something else was going on in my heart. There were other emotions, long suppressed, gurgling to the surface and they wouldn't go away. I was becoming more and more averse to violence. Watching TV or movies in which there might be violence was out of the question. I could remember only too clearly that while on TDY at Fort Knox not long before retirement I went to the post theater to see "The Deer Hunters." I couldn't handle it and walked out in the middle of it confused over what was going on inside of me. I was becoming more and more sensitive to grief.

The ghosts of remembering the funerals at West Point and the Shine episode seemed always ready to call me back to that which I did not want to recall. Whenever I saw a man with a "Vietnam Vet" baseball cap walking around on crutches missing one of his legs, I would quickly run off and hide someplace where people would not see or hear me sobbing out of control. My wife had to protect me from seeing those disabled vets and from other people who started asking too many questions about Vietnam. I was becoming more and more aware of my sinfulness and how much I had hurt the One Who died for me. I didn't want to go back to West Point to any reunions. Too many demons were waiting for me there. I became totally separated from West Point, my classmates, and the Army. I kept in touch with only two of my classmates, and them because when together we would talk about hunting and other things, not the Army.

Somewhere along the line I learned why Jon's mother was so immediately sensitive to the appearance of a uniformed Army officer at her door. Her father had served with the American Expeditionary Force in World War I, and her husband in the European theater of World War II. Al had been wounded and medically evacuated from Vietnam some years before. She knew Army officers ringing your doorbell in the middle of the day were not bearing good news. That only added to my heartache for the Shines. And I could never escape the painful thoughts of the price the Shine clan paid for their patriotism. Tears were never more than a few seconds away.

But still another thing was going on in my heart. I began to see that God had a huge and wonderful plan for ME when He chose me to carry the bad news to the Shines. Before I even knew what being a Christian was all about I saw it in action. Although beyond the red eyes and wet cheeks I cannot clearly remember Jon's mom and dad's faces when they reappeared after their long retreat behind the closed doors, I can still remember their demeanor. They were confident. They had direction. Heartbroken, yes. But they had something I didn't. So did everybody else named Shine. And I am persuaded that is why I got involved in Bible studies before leaving West Point for Vietnam again. And why in July of 1971 in Lawton, Oklahoma, I received the forgiveness of my sins. And why now I can say that Jesus is my Lord and Savior and why all of my dreams about what life was supposed to be like when I reached 68 years of age are coming true.

This has not been a tearless endeavor but is wonderfully liberating. I don't ever have to tell people this story again. I can send it to them and it is significantly less painful to me that way.

But those grievous memories are still waiting in the wings for a moment of weakness. Fading a little each year, perhaps, but way too slowly.

## Testimonials of Jon Shine's Friends, Family, and Classmates

### Colonel (retired) Al Shine
Jon's older brother, a career Army officer and after retirement, ten years as Commandant of Culver Military Academy, Culver, Indiana:

> *What has endured to me from Jon are at least two things: First — his example. What Jon had that*

*neither Tony nor I had was balance...As a Christian I'm sure the Holy Spirit further refined these natural qualities of his personality and used them in the lives of others. His balance and outgoing personality are a challenge and example to me. Second—the sense that I have a special obligation to live out faithfully my calling in this world because of the three brothers, I am the one God has chosen to have a longer life on this earth. I know that Jon is among the "great cloud of witnesses" cheering me on, and knowing this I am both challenged and strengthened to seek to be the man God wants me to be and serve Him and those He would have me serve in His strength.*

## Robert M. Kimmitt

Classmate and friend of Jon, former Under Secretary of State for Political Affairs, former U.S. Ambassador to Germany and Major General, U.S. Army Reserves in an e-mail to Barry Willey:

*Thanks for this opportunity to remember a friend and first-class human being...we got to know each other working on various Corps-wide activities and via mutual friends. One such friend was Guy Hester... who was also killed in Vietnam [Authors' note: seven days before Jon]. Jon and Guy were devout Christians, but Christians who were not judgmental about others. They influenced others by example, not sermonizing.*

> *I believe Jon's and Guy's widows were burying one of them...when word of the other's death arrived.* [Author's note: Gail Shine rushed to be with Guy's wife – a total stranger – when she heard that Guy had been killed in action. While Gail was at the funeral, incredibly, Guy's wife found out about Jon's death and informed Gail. They comforted one another.]
>
> *When I think of what a West Point cadet should be, I think of Jon Shine. When I think of what a young officer should be, I think of Jon Shine. When I think of what a human being should be, I think of Jon Shine. When I think of Jon Shine, I cry and smile; cry because I miss him, smile because he is now where he was always destined to be.*

## Colonel (Ret.) Mike Tesdahl

USMA Class of 1969, classmate of Jon Shine and former head of the Officers' Christian Fellowship ministry at West Point, in a correspondence to Barry Willey:

> *The Jonathan C. Shine Memorial Award, traditionally a leather-bound study Bible engraved with the recipient's name, is still presented annually during the Protestant Baccalaureate service. The award is presented to the Cadet-in-Charge of OCF at West Point. This cadet is chosen on the basis of his/her demonstrated capacity for Christian leadership and service within OCF and the Corps. Jon's contribution to the spiritual*

development of his contemporaries is reiterated annually in preparation for this award ceremony.

Recipients of the Award include Bryan Groves, Mike Stone, Marie (Roush) Hatch, and Riley Post. Jon is also a recurring topic of discussion during the annual West Point-specific Rocky Mountain High rotation at Spring Canyon, Colorado, where Fort Shine is named after him. In particular, we recount the anecdote that Jon was led to the Lord through the witness of a "barracks police" — a custodian — at the Gillis Field House, and make the point that none of us knows the impact of our daily witness on those around us.

## Dr. Art Hill

Professor in the Operations and Management Science Department of the Curtis School of Management, University of Minnesota, and a former Military Academy classmate of Barry Willey before leaving the Academy for Indiana University:

> *I only met Jon Shine once or twice. After I accepted Christ in the gym when you and Paul (Stanley) were there, you guys invited me to Paul's house once or twice before I left the Army. I clearly remember you praying – and grunting – with more passion and fervor than I had ever seen. You guys REALLY prayed to God. This was none of the "let's go through the motions" prayer. You, Paul, and that group of guys really had an impact on my life. You guys loved me and valued me at a time in my life when I really needed it. That is all that I remember about Jon or that group of guys.*

... and in a letter from Art, with a letterhead from Campus Crusade for Christ ministries, dated Fall 1971, Art shares with friends the details of his changed life:

> *After high school I went to the United States Military Academy at West Point. One of my friends came to me and talked about a <u>personal</u> relationship with Jesus Christ. He shared the gospel with me many times–and then one day I came to*

understand. At that time I received God's free gift, Jesus Christ, as my Lord and my Savior.

… and Art pens a personal note to Barry at the bottom of the form letter:

> Brother—I want you to know that I really appreciate you! I am now a spiritual grandfather about ten times over—which makes you a great granddad….

## Mike Hulten
West Point classmate and company mate, in a letter to Al Shine:

> I have been impressed for some time to drop a note to your folks to let them know what a positive example Jon was to me, and to many others as well. I just want to let all of you in the family know what a special person Jon was. To be honest, he and I were not "close," although in the same company. I did not have the privilege of rooming with him or anything special like that. But perhaps that is a testimony of Jon's greatness. He touched all of us with who he was and what he stood for. He has always been something of a hero to me.
> Like many of us there at West Point, I had a great love for my country and a great desire to serve…but I was not from a family that had a

*strong spiritual background. Jon had something different about him, a special quality. I did not know what it was at the time, just that he was always happy, always positive. He set a strong example; one of integrity, thoughtfulness, and firmness of character that set him apart, even among as strong a cast of characters as we enjoyed there.*

*It is only in the past few years, as insight and priorities have come clearer to me, that I can see what it was about Jon that made him special... his example and memory are strong. There are undoubtedly others who have been similarly touched. Example is always such a powerful teacher.*

*Thank you for making him into what he is, and for the example that he — and by extension, you — have shared with us. May you know that his service, in so many areas, is still felt and appreciated...I know that there will be another "roll call," another opportunity for us to be together. I look forward to the time — as I know you do — of shaking his hand in that always firm grip, seeing again that ever-present smile, and expressing personally my appreciation to him for his service.*

## Dave Jamison
Jon's West Point roommate:

> *Jonathan C. Shine was unquestionably brilliant and talented, but his influence was due to his unique and exceptional character. Although I believed myself a Christian when I arrived at (West Point), I learned so much from Jonathan about commitment to Christ that it is hard to explain. Much of my convictions developed over the past thirty years have a parallel in something I observed in Jon. I emphasize this aspect of my memories first because it was Jon's deep faith that made his many other talents so exceptional. In success, he made it clear that Christ was with him in everything he did. From the day I met him until our graduation, Jon was an inspiration for me to strive to be a better Christian.*

## Colonel (Retired) Jeff Jones
G-1 company mate, friend, and former United States Army Defense Attaché, Paris:

> *When they played taps at the funeral, all of us were, once again, reminded of the price of war – on the soldier, sailor, airman or Marine – on the widows, children, families, loved ones and friends... perhaps on the nation and its population – indeed on the Vietnamese. What were the chances that*

*those of us still cadets that day suffer the same fate? Would we be the kind of leaders who were respected, loved and followed by our subordinates, as Jon Shine had been? Would we measure up to his standards of Duty, Honor, Country? Of an enduring, albeit all too short, commitment to his soldiers?*

*That was my memory of Jon Shine — intellectually gifted, competent, committed, compassionate, dedicated, selfless, a leader who exuded all those qualities and attributes of the ideal West Pointer upon whom soldiers, units, the Army, and the nation could depend. Someone with a bright future who would unquestionably be one of the shining stars of tomorrow — only tomorrow didn't come.*

*Nonetheless, he did indeed shine brightly when he was with us. He inspired us by his example and his personal standards of right and wrong, of duty to be performed, of country to be ever armed. And suddenly he was gone, and at peace, but with an enduring legacy for us all — soldier, teacher, leader, advocate, defender, protector, big brother, parent, chaplain, friend. He did shine. And his memory and legacy still shine today. The most important thing a human being can do is to inspire. Jon Shine did that.*

## William J. Bahr
Classmate of Jon:

> *As I write you, I have beside me a profile honoring Jon...You may rightly infer that I thought much of him. Indeed, I'm sure this positive feeling must have subconsciously played a part in the naming of my own son, Jonathan. Upon graduation from West Point in 1969, I left with many fond memories, not the least of which was a picture of what a perfect cadet might look like. That mind's-eye picture was of Jon Shine.*
>
> *In 1989, after I read new information about him in the profile clipping, that picture was still Jon Shine. I am sure as your book gathers more about him, the picture of this "finest" cadet, this remarkable person in every sense of the word, will become clearer. Thank you for honoring his memory.*

## Colonel (Retired) Ralph Puckett
Columbus, GA., former Regimental Tactical Officer at West Point and awarded two Distinguished Service Crosses for heroism in the Korean and Vietnam conflicts, in a letter to his granddaughter, Lauren, before her school group traveled to Washington, D.C.:

> *While you are at the Wall [Vietnam Memorial], you may have the time to make a "rubbing" of one*

*of the names. I suggest that you look for the name of Jonathan Cameron Shine. Jon means something to me and to your mother. Jon was your mother's Sunday school teacher when we were at West Point. Your mother was a teenager then. She thought Jon was wonderful — and he was. Jon was one of the most outstanding young men I have ever known. He was very bright, standing very high in his class academically. He was ranked very high militarily. He was a Cadet Battalion Commander in my regiment. He was a great athlete, earning honors as the Eastern Intercollegiate High Bar Champion in gymnastics. More important than all those honors was Jon's character. He epitomized integrity. I hope that someday you find a man like Jon, grow to love him, and marry him. If you have any sons, I hope you will raise them to be like Jon. If you do, the world will be a better place.*

[Author's note: Al Shine shared with me that when he was at Fort Benning in 1982-84, commanding an infantry training battalion, he went to a Ranger graduation where Colonel Puckett was speaking. After the ceremony he introduced himself to Colonel Puckett. Turning to those around them, Colonel Puckett's words were succinct and to the point, "If this man is half the man his brother was, he's the best man here."]

## LTC (Ret.) Marv Wooten

One significant group of young people whose lives have been powerfully affected by Jon are the cadets involved in the Officers' Christian Fellowship (OCF) ministry at West

Point. LTC (Ret.) Marv Wooten led that ministry for many years and shared insights from his experiences:

> *We created the Jonathan Cameron Shine memorial award in 1992 in order to honor and recognize the contribution to the Corps of Cadets by the Cadet-in Charge (CIC) of the Officers' Christian Fellowship — United States Corps of Cadets. OCF at USMA is sponsored by the USMA Chaplain and is a part of the officially sanctioned, authorized cadet religious activities under the Director of Cadet Activities. While searching for a name/namesake for the award, we looked at several possibilities. But we chose Jon because of his remarkable Christian leadership both as an officer and particularly while a cadet.*
>
> *Your writing about his influence in your life was a big factor in this as well. I was struck by how, at a time when being "religious" was especially not cool, Jon practiced his faith with fervent devotion... also, in how he manifested that faith in his witness to fellow cadets. That he was taken from us at a comparatively early age while offering such a wonderful role model enables cadets to identify all the more with him and the award. As cadets become aware of Jon's life as a cadet and young officer, I would hope they are inspired. That was one of the main reasons we chose to name the award after him.*
>
> *The CIC's who have received the award have each been exemplary in their Christian lives. The*

*first award was presented to Cadet Mason Crow, '93. Notably, the 1997 selectee, Dan Hart, was also First Captain for his class (the most senior cadet in rank — the top cadet in the Corps of Cadets). Jay Bartholomees, the '95 awardee, served as a 1LT in the Ranger Battalion at Fort Lewis (Washington state).*

## Army Captain Gwynn Vaughan

When Paul Stanley left the Army, another faithful Christian member of the Academy faculty, Army Captain Gwynn Vaughan, took up the leadership reigns and continued to guide many of the young cadets seeking spiritual guidance. Gwynn shared an anecdote with me that clearly put Jon's death into the proper perspective:

> *I have compared Jon with (a Class of '56 All American football player) Ray Smith [author's note–name changed to protect family's privacy] who also had great credentials and was killed as a brigade S-3 operations officer in Vietnam. He was my boss in Vietnam and plebe football coach at USMA in 1960. To my knowledge he had no commitment to Jesus, but great worldly potential for future Army leadership.*
>
> *Then I compared their lives to the Army-Navy football game in 1963 when Army was on the one-yard line, down 21-15, with 13 seconds to play. With all that was riding on the outcome — national ranking (Navy was #2, Army #17), the Lambert*

*Barry E. Willey*

*Trophy for best team in the East, the winner to the Cotton Bowl to play #1 Texas, and most importantly beating Navy — we couldn't get the last play off and were left on the one-yard line — much like Ray Smith. Jon, however, is in victory land with Jesus...and through his influence he has shown the way for others to follow.*

# Acknowledgements

This book has been a 36-year effort. That means there are many who have touched my life in many ways and enabled me to put it all together.

Most importantly, my Lord and Savior, Jesus. I have prayed throughout this process that God would stop the progress if I intended in any way to do *anything* but *exalt Him and build His kingdom through this book.* He has seen it through to completion and for that I am very grateful.

My mother and father, Mary and Carroll Willey, raised me in a loving and nurturing environment, planting the seed of the Christian faith in my life early and allowing it to germinate and grow into a full-blown faith walk. Dad loved Mom… faithfully and consistently. That act did more for me than anything else. It showed me what the Apostle Paul meant when he said, "Husbands, love your wives, just as Christ loved the church and gave himself up for her." (Ephesians 5:25)

COL (Ret.) Al Shine and his dear wife, Sandra, have been role models for Barb and me since we met them in 1972. They showed us, before we were married, what life could be like as a Christian married couple and family in the Army. Al was an amazing Army officer and gave me a glimpse of what a professional standard-bearer for the Army, the nation, and the Lord looked like. And, of course, Al's help, support, and insights into life in the Shine family, and particularly Jon's life, were invaluable and allowed me to get to really know sides of Jon's life I otherwise wouldn't have known. Their friendship has been... and is... priceless.

Dr. Jim Blackwell, USMA, Class of 1974, greatly helped me in the initial stages of the very first manuscript. He offered to help, as he had published a couple books, and drafted several book proposals when we were looking for a publisher. He advised me on the drafting of the first manuscript. His help was invaluable in inspiring and motivating me to press on with this effort.

Many dear brothers in Christ invested their lives in me. I cannot mention them all. A few I want to recognize are Paul Stanley, Gwynn Vaughan, Dr. John George, Pastor (Dr.) Jack Elwood, Pastor (Dr.) Tim Cole, Chaplain Earl Andrews, and my current boss, LTG (Ret.) Bruce Fister. All have powerfully exemplified another of Paul's classic statements—"We loved you so much that we were delighted to share with you not only the gospel of God but our lives as well, because you had become so dear to us" (1Thessalonians 2:8). Their investment is reaping eternal dividends.

Most dear to me are my wife and children—my dear Barb, Rachael, and Jonathan, Jonathan's precious wife, Jamie, and our two amazing grandsons, David and Michael. Their personal faith stories at the end of this book speak for themselves. Rachael and Jonathan were the best possible Army children I could have hoped for and make our lives so full and rich. They are faithfully walking with Christ and bring untold joy into our lives. Barb is the truest love of my life, my dearest friend, and the ultimate life partner. No man could ask for or dream of a more wonderful mate. Thank you, my dearest family.

Any omissions, errors, or oversights in this account are my responsibility alone. I am eternally grateful for the opportunity Officers' Christian Fellowship, and particularly Mike Edwards, Director of Communication, and Barb Beyer, his faithful assistant and primary editor of this publication, have given me to share this amazing story of a life—brief, but well-lived—with the world. I pray it will inspire and motivate those who read it to exalt Jesus Christ, invest in others, build His Kingdom… and leave a legacy.

Barry E. Willey, COL (Ret.), U.S. Army
West Point, New York
August 2007

# Biographical Information Colonel Barry Willey, U.S. Army (Retired)

Barry was born in Batesville, Indiana, in July of 1950. He moved with his family to the Indianapolis area when he was four years of age. In 1968, Barry graduated from North Central High School and headed to West Point, New York, attending the U.S. Military Academy. He graduated in 1972 as a Second Lieutenant of Infantry in the U.S. Army.

He married Barb in 1974 and they became the parents of Rachael and Jonathan. As a family, they traveled throughout the United States and the world serving the Lord, the U.S. Army, and the nation. He spent much of his military service in infantry units (82nd Airborne Division, 24th Infantry Division, The Ranger Training Brigade, among others). As a paratrooper and Army Ranger, he was privileged to serve with some of the finest service men and women in our country.

In 1982, Barry attended Indiana University to earn a Master's Degree in Journalism and go on to spend about 10 years in the Army as a Public Affairs Officer, helping tell the Army's great story. His deployments, as both an infantryman and public affairs officer, included combat operations on the Island of Grenada in 1983, combat operations during Desert Shield and Desert Storm in 1991, and stability operations in Haiti in 1994. During his time in Haiti, he was the spokesman for the U.S. military forces, holding more than 100 press conferences and interviews with the world's media, including many live TV events.

After 29 years of active duty—and 33 years affiliated with Officers' Christian Fellowship—he retired from the Army in 2001 and became a strategic communications consultant with a Washington, D.C.-based firm, Booz Allen Hamilton.

Barb and Barry were excited, honored, and humbled to be selected as the field staff representatives for Officers' Christian Fellowship at West Point in January 2005. Before going back to West Point, Barry had led dozens of OCF Bible studies and taught Sunday school in military chapels throughout three decades in the Army, while personally discipling nearly 60 men in their Christian faith-walk. Barb was the Director of Children's Ministry at Burke Community Church in Burke, Virginia, and Barry was selected to be an elder and was serving as a three-and four-year-old Sunday school teacher and a men's discipleship trainer, when they were called to the OCF West Point ministry.

Following the West Point campus ministry, Barry returned to Booz Allen Hamilton as an Associate in strategic communications. He also has most recently worked with two U.S. Government organizations—The National Commission on the Structure of the Air Force, and The United States Vietnam War 50th Anniversary Commemoration, providing public affairs and communication support.

Barb is now a real estate agent and Barry is a part-time communication consultant. They are the proud grandparents of David Ian and Michael Edward, who were born to their son and daughter-in-law, Jonathan and Jamie, who live in Cooper City, Florida, where Jonathan is Director of Photography and Creative at the Miami Dolphins football team and Jamie is a trained youth leader in the United Methodist Church. Barb and Barry's daughter, Rachael, was an elementary school teacher for eight years and is now a nanny in St. Petersburg, Florida.

# Resources

Additional books that the author has used that will assist in preparation for spiritual mentoring discipleship and investment:

- *With Christ in the School of Disciple Building* by Carl Wilson (1976)

- *Connecting: The Mentoring Relationships You Need to Succeed in Life* by Paul Stanley and Robert Clinton (1992)

- *The Master Plan of Evangelism* by Dr. Robert Coleman (1987)

- *The Master Plan of Discipleship* by Dr. Robert Coleman (1997)

- *The Mind of the Master* by Dr. Robert Coleman (2000)

- *The Complete C.S. Lewis Signature Classics* (2002)

- *Questioning Evangelism: Engaging People's Hearts the Way Jesus Did* by Randy Newman (2004)

*Barry E. Willey*

- *Resilient Warriors* by Maj Gen (Ret) Robert Dees (2011)

- *The Invested Life: Making Disciples of All Nations One Person at a Time* by Joel Rosenberg (2012)

# Products and Services

## A Heart for Presentations

Barry Willey has been discipling believers, investing his life in theirs, since 1971. His heart and passion for mentoring others has guided him, along with the Holy Spirit, to write both *Out of the Valley* and *Extreme Investing*. He is available to organizations, churches, and small groups to share this passionate love for investing in others.

## Offerings

- *Out of the Valley*: An amazing life story that can help you make good choices… and leave an eternal legacy (Creative Team Publishing, 2015)

- *Extreme Investing: Changing the World One Believer at a Time* (Creative Team Publishing, 2015)

- Teaching, training, mentoring sessions

- Camp and conference speaking

## Business Presentations
- Investing that Matters in Employees
- Investing that Matters in Leaders and Managers
- Investing that Matters in Families

## Military Group Presentations
- Investing that Matters in Small Unit Leaders
- Investing that Matters in Senior Leadership
- Investing that Matters in Key Staff

# Investing That Matters Website

www.InvestingThatMatters.com

To contact the author or locate author's books:

- www.barrywilley.com

- www.outofthevalleybook.com

- www.extremeinvestingbook.com

- www.investingthatmatters.com

www.ingramcontent.com/pod-product-compliance
Lightning Source LLC
Chambersburg PA
CBHW022227010526
44113CB00033B/644